AF762690

CATALOGUE OF WHITE DWARFS

CATALOGUE OF WHITE DWARFS

by

A. G. AGAYEV, O. H. GUSEINOV, and H. I. NOVRUZOVA

Reprinted from

Astrophysics and Space Science, Vol. 81, Nos. 1/2

D. Reidel Publishing Company

P.O. Box 17, 3300 AA Dordrecht, Holland

Order Ref. No. 90-277-9054-1

All Rights Reserved
Copyright © 1982 by D. Reidel Publishing Company, Dordrecht, Holland
No part of the material protected by this copyright notice may be reproduced or utilized in any form or by any means, electronic or mechanical, including photocopying, recording or by any informational storage and retrieval system, without written permission from the copyright owner

Printed in The Netherlands

CATALOGUE OF WHITE DWARFS*

A. G. AGAYEV, O. H. GUSEINOV and H. I. NOVRUZOVA
Shemakha Astrophysical Observatory, Baku, U.S.S.R.

(Received 15 April, 1980)

Up until now there has not been available a uniform catalogue of white dwarfs which contains all the available principle data. Previously, various authors, including Eggen and Greenstein (1965, 1967), Luyten (1962), Greenstein (1972), and Greenstein *et al.* (1976) have presented restricted lists of the number of white dwarfs and their individual characters, etc.

All the present data (spectroscopic, photometric, kinematic) for 488 white dwarfs have been collected together by us. These data were systematized and revised by new information. The catalogue contains the following information:

Column (1)	number,
Column (2)	number by EG or name of star, source of identification,
Columns (3, 4)	coordinates for 1950 yr,
Column (5)	spectral type,
Column (6)	effective temperature or $\theta = 5040/T_{\text{eff}}$; in some cases calculated for two variants of luminosity,
Column (7)	M absolute magnitude,
Column (8)	R radius of star,
Column (10)	m_{pg} photographic star magnitude,
Columns (11–18)	colours in UBV,
Column (20)	common proper motion of the star or its components,
Column (21)	θ positional angle,
Columns (22, 23)	the values of (U, V, W) in km s^{-1} from the proper motions, for the possible luminosities (from the assumption that star belongs on the upper or lower sequences, respectively),
Column (24)	the percentages of the radial velocity that should be added, respectively, to U, V, and W,
Column (25)	the radial velocity with correction relatively to the Sun and to red shift in km s^{-1},
Column (26)	V_t km s^{-1} transverse or tangential velocity,
Columns (27, 28)	trigonometric and photometric parallaxes π,
Columns (9, 19, 29)	references to utilized literature.

In the catalogue the following signs are also used:
W and $W_{0.5}$ – equivalent widths and half-widths of the lines in Å, respectively.

* Astrophysics and Space Science Review paper.

Astrophysics and Space Science **81** (1982) 5–84. 0004–640X/82/0811–0005$12.00.
Copyright © 1982 *by D. Reidel Publishing Co., Dordrecht, Holland, and Boston, U.S.A.*

In colours $(B - V) = m_{2.12} - m_{1.44}$ (according to Nesterenko and Chuprin, 1976).
$\{B - V\}$ = rough data.

Δm_{UBV} – length of perpendiculars (in star magnitudes), formed from the points, which correspond to objects on $U - B$, $B - V$-diagram, on black-body curve. The sign '−' or '+' means, that the object is higher or lower than black-body curve,

Δm_{pg} – modulus of maximal difference of photographic magnitudes for a given object.

Notes

1. $B - m_{pg} = 0.9$, $\Delta m_{UBV} = -0\overset{m}{.}06$.
4. Binary. Component G 130-50: $V_E = 13\overset{m}{.}21$, $V - R = 0\overset{m}{.}99$, $R - I = 0\overset{m}{.}87$, $B - V = 1\overset{m}{.}41$, $U - B = 1\overset{m}{.}11$, $M_v = 9\overset{m}{.}7$.
6. Mass of white dwarf = 0.75 $\mathcal{M}_\odot$ [50].
11. Mass of white dwarf = 0.96 $\mathcal{M}_\odot$ [50].
 $\Delta M = +0\overset{m}{.}32$, $U - V = 0\overset{m}{.}29$ [35].
8. Possibly binary with G 171-62? Mass of white dwarf = $0.25\overset{m}{.}\ \mathcal{M}_\odot$ [50].
14. Binary with GD 218-7. $M_{pg} = 16\overset{m}{.}0$ – for component. Sep. between components $\rho = 11''$, $\theta = 246°$.
15. $B - m_{pg} = 0\overset{m}{.}2$, $\Delta m_{UBV} = -0\overset{m}{.}13$, $\Delta(U - B) = 0\overset{m}{.}18$, probably, variability. There is probability of relativistic object [5].
16. Magnetic white dwarf. Magnetic field 18-20 (MG). Circular polarization 0.3%.
22. $W = 19$ Å, $W_{0.5} = 40$ Å. Possibly, the member of Hyades cluster.
23. $W(\mathrm{H}\gamma) = 17.1$ Å, $W_{0.5} = 23$ Å [10]; $U - V = 0\overset{m}{.}45$, $\Delta M = +0\overset{m}{.}93$ [35]. White dwarf mass = 0.36 $\mathcal{M}_\odot$ [50].
27. Binary.
29. Luminosity $L = 5 \times 10^{30}$ erg s^{-1}. $B - m_{pg} = -0\overset{m}{.}05$, $\mathcal{D}m_{UBV} = -0\overset{m}{.}06$ [5]. There is circular polarization $g < 20\%$.
32. $W = 28$ Å, $W_{0.5} = 33$ Å, $W' = 10$ Å, $W'_{0.5} = 48$ Å [1].
 White dwarf mass = 0.20 $\mathcal{M}_\odot$ [50].
34. Mass of white dwarf = 0.05 $\mathcal{M}_\odot$ [50].
36. $W(\mathrm{H}\gamma) = 6.0$ Å, $W_{0.5} = 7$ Å [10]; $U - V = 0\overset{m}{.}16$, $M = -1\overset{m}{.}84$ [35], peculiarities in the spectra.
38. Binary with G 34-48. Component's characteristics: $V_E = 13\overset{m}{.}13$, $V - R = 0\overset{m}{.}97$, $R - I = 0\overset{m}{.}82$, $B - V = 1\overset{m}{.}45$.
 Spectral type Sd M3 the distance $\rho = 24''$, $\theta = 60°$ [5].
 $\Delta m_{pg} = 1\overset{m}{.}1$, probably, variability in the brightness. Luminosity $L = 2 \times 10^{30}$ erg s^{-1}. Wide pair with the suspicion on relativistic object [5] $B - m_{pg} = 1\overset{m}{.}2$; $\Delta m_{UBV} = 0\overset{m}{.}19$ [5].
40. Possibly nondegenerate star.
44. Binary, with pair with G 71-B5A. The component's characteristics: $V_E = 14\overset{m}{.}25$, $V - R = 1\overset{m}{.}04$, $R - I = 0\overset{m}{.}905$, $B - V = 1\overset{m}{.}43$, $U - B = 1\overset{m}{.}09$, $M_v = 9\overset{m}{.}9$ [10].
46. Mass of white dwarf = 0.50 $\mathcal{M}_\odot$ [50].
47. Mass of white dwarf = 0.54 $\mathcal{M}_\odot$ [50], $\Delta m = -0\overset{m}{.}12$ [35].
49. Mass of white dwarf = 0.70 $\mathcal{M}_\odot$ [50], $\Delta m_{UBV} = 0\overset{m}{.}04$ [5], $\Delta m = -0\overset{m}{.}53$ [35].
51. Binary with G 94-B5A. The component's characteristics: $V_E = 8\overset{m}{.}30$, $B - V = 0\overset{m}{.}61$, $U - B = 0\overset{m}{.}10$, $M_v = 5\overset{m}{.}0$, spectr. type dG. Mass of white dwarf = 1.09 $\mathcal{M}_\odot$ [50], mass of component ≈ 1 $\mathcal{M}_\odot$.
55. Mass of white dwarf = 0.11 $\mathcal{M}_\odot$ [50]. Possibly, old Nova.
59. Mass of white dwarf = 0.68 $\mathcal{M}_\odot$ [50].
64. Binary, has component at the distance $14''8$.
65. Binary, in pair with G 5-27. Component's characteristics: $V_E = 7\overset{m}{.}41$, $V - R = 0\overset{m}{.}36$, $R - I = 0\overset{m}{.}29$, $B - V = 0\overset{m}{.}78$, $U - B = 0\overset{m}{.}35$, $M_v = 6\overset{m}{.}0$, spectr. type G 5 [10]. The star itself is an old object of Population type I.
68. Binary with G 221-11. Components are at the distances $\rho = 13''$, $\theta = 2°$. In this case there are indications on orbital motion, as positional angle θ has decreased on 5° during the last 25 yr. [5].
69. Binary with L 587-77B. White dwarf, evidently has $M_v = 11^m$ and is situated on the upper branch of H-R diagram. Component-spectral type Sd M3 and has very slight lines Ca II and Al I and strong Ca I and TiO.

70. $W(\mathrm{H}\gamma) = 30.6$ Å, $W_{0.5} = 35$ Å.
72. Peculiarities in λ 4670; $\Delta M = -0^{m}.77$.
76. Proposal initial mass of white dwarf was $\approx 5\,\mathcal{M}_\odot$ (39). Final mass $\approx 0.9\,\mathcal{M}_\odot$ – by lines Schulz (1976) and $\approx 0.52\,\mathcal{M}_\odot$ by red shift in [4]; $m = 0.96\,\mathcal{M}_\odot$ by [50] $\Delta m = 0^{m}.59$. Possibly member of Pleiades.
77. $m = 0.60\,\mathcal{M}_\odot$ [50].
Initial mass of the star, progenitor mass $m = 2.9\,\mathcal{M}_\odot$ [39]. Belongs to Hyades cluster.
78. Spectra are very peculiar, Ca II lines are sharp, Ca I are wide and slight. There are blends Fe I in UV. This star may be Sd Mp without metal and TiO. Proper motion of object is approximating to W 219 and M_v is calculated from assumption, that it belongs to group W 219 [10]. Possibly nondegenerate star.
79. $m = 0.70\,\mathcal{M}_\odot$, in Hyades? [39].
80. In Hyades.
$m = 0.92\,\mathcal{M}_\odot$ [50].
Progenitor mass $\approx 4\,\mathcal{M}_\odot$.
Present masses: $0.95 - 0.8\,\mathcal{M}_\odot$, $0.79 - 0.63\,\mathcal{M}_\odot$ [39].
81. In Hyades.
84. In Hyades. Possibly subdwarf with slight lines.
85. $m = 0.56\,\mathcal{M}_\odot$ [50]. $m = 0.43 \pm 0.2\,\mathcal{M}_\odot$ [39].
Triple system.
90. $m = 0.66\,\mathcal{M}_\odot$; progenitor mass $\approx 2.85\,\mathcal{M}_\odot$ [39].
91. There is circular polarization $g < 0.22\%$.
$B - m_{pg} = 0^{m}.3$.
93. Wide binary with G 175-34A.
Component's characteristics: $V_E = 11^{m}.10$, $B - V = 1^{m}.62$, $U - B = 1^{m}.14$, $M_v = 12^{m}.4$. Spectr. type of component Sd M4 [10]. Sep. $\rho = 6''.8$, $\theta = 145°$. The shortest period of rotation among close binaries – systems takes place in this pair. Data on ρ and θ are given for 1908 yr. In 1971 yr.: $\rho = 6''.5$ and $\theta = 82°$. For a year θ is changing more than by 1°. Rotation period of such system 300–350 yr. Luminosity $L = 10^{30}$ erg s^{-1}, there is circular polarization of $g < 0.26\%$.
94. In Hyades; possibly, binary w.d.; composite spectra.
95. In Hyades; $m = 0.71\,\mathcal{M}_\odot$ [50], progenitor mass $\approx 2.6\,\mathcal{M}_\odot$ [39].
96. Binary with G 39-28. Component's characteristics: $V_E = +8^{m}.2$, $V - R = 0^{m}.53$, $R - I = 0^{m}.51$, $B - V = 1^{m}.12$, $U - B = 0^{m}.88$, $M_v = 7^{m}.4$. Spectr. type dK pe. Radial velocity of the component, which is peculiar emission star is given in data. System is external member of Hyades. It is situated at the distance 16 pc from Sun [10]. There is circular polarization $g < 30\%$, $\rho = 128''$, $\theta = 338°$ [5]. Luminosity $L = 4 \times 10^{29}$ erg s^{-1}. By [50] $m = 0.72\,\mathcal{M}_\odot$.
100. In Hyades cluster; $W(\mathrm{H}\gamma) = 20$ Å. $W_{95} = 33$ Å [10]. Progenitor mass $\approx 2.5\,\mathcal{M}_\odot$ [39]. By [50] $m = 0.80\,\mathcal{M}_\odot$.
105. Binary G 191-B2A and G 191-B2B.
Component Sd K4p, and possibly white dwarf $m_{\mathrm{w.d.}} = 0.29\,\mathcal{M}_\odot$ [50].
107. Binary with G 86-B1A. Component characteristics: $V_E = 14^{m}.20$, $V - R = 1^{m}.00$, $R - I = 1^{m}.03$, $B - V = 1^{m}.66$, $U - B = 1^{m}.41$, $M_v = 10^{m}.7$.
$W(\mathrm{H}\gamma) = (19)$, $W_{0.5} = (37)$; by [50] $m = 0.49\,\mathcal{M}_\odot$.
111. Magnetic field 4 (MG).
Circular polarization 0.8%, linear polarization $\leq 0.1\%$.
113. Binary with GC 7413. Component's characteristics: $V_E = 7^{m}.76$, $B - V = 0^{m}.58$, $U - B = 0^{m}.08$, $M_v = 4^{m}.8$. Spectr. type of component (F8). $\rho = 91''$, $\theta = 47°$ [5] – wide pair with the suspicion of a relativistic object. Luminosity $L = 10^{31}$ erg s^{-1}, $R = 10^9$ cm. There is circular polarization $g < 0.22\%$.
114. Mass of white dwarf $0.57\,\mathcal{M}_\odot$ [50].
115. Polarization in ultraviolet $g = 0.45\%$, in near infrared region $g = 0.30\%$; linear polarization $< 0.2\%$ [5].
Star probably variable white dwarf BS. Mass $= 0.64\,\mathcal{M}_\odot$ [50].
120. There is no visual helium lines: $W(\mathrm{H}\gamma) = 17.5$ Å, $W_{0.5} = 29$ Å [10] $m = 0.70\,\mathcal{M}_\odot$ [50].
121. Common proper motion with component HR 2274 at the distance 41″.
124. Binary with G 105-B2A. Component's characteristics: $V_E = 13^{m}.25$, $V - R = 1^{m}.05$, $R - I = 0^{m}.905$, $B - V = 1^{m}.48$, $M_v = 9^{m}.9$ [10]. Component of spectr. type M2. Spectra $M = 13^{m}.2$ shows, that star is one of the faintest stars of low branch of the sequence.
129. Sirius C. $m = 1.03\,\mathcal{M}_\odot$ [50].
$W(\mathrm{H}\gamma) = (22)$ Å, $W_{0.5}(\mathrm{H}\gamma) = (32)$ Å [10].
130. Mass of white dwarf $= 0.83\,\mathcal{M}_\odot$ [50], $W(\mathrm{H}\gamma) = 28.8$ Å.
Mass of white dwarf $= 0.23\,\mathcal{M}_\odot$ [48], $W_{0.5}(\mathrm{H}\gamma) = 40$ Å [10].

137. Binary with G 87-28. Component's characteristics: $V_E = 14^m.60$, $V - R = 1^m.34$, $R - I = 1^m.25$, $B - V = 1^m.68$, $U - B = 1^m.21$, $M_v = 12^m.2$; spectr. type Sd M6. Slight line λ 4670 [10].
140. Possible member of Hyades; there are wide lines in spectra; there is polarization; $g < 0.20\%$ [5].
142. Binary with G 107-69. Component's characteristics: $V_E = 13^m.52$, $B - V = 1^m.69$, $U - B = 1^m.15$, $M_v = 13^m.5$, spectr. type Sd M5. There are no visual lines from violet to λ 6700 Å. Sep. components $\rho = 105''$, $\theta = 155°$ [5].
Mass = 0.37 $\mathcal{M}_\odot$ [50]. Luminosity $L = 2 \times 10^{29}$ erg s^{-1}. Circular polarization power $g < 0.20\%$.
146. Procyon. Binary.
147. Possibly, binary with component $16^m.0$ and sep. 21″, $W(\mathrm{H}\gamma) = 5.8$ Å, $W_{0.5} = 10$ Å [10], $\Delta m = 0^m.34$ [35].
153. Mass of white dwarf = $0^m.76$ $\mathcal{M}_\odot$ [50], $\Delta m = 0^m.06$ [35].
154. Circular polarization power $g < 0.08\%$ [5]; Luminosity $L = 3 \times 10^{29}$ erg s^{-1}. Possibly variable white dwarf. It is suspected of being a relativistic object by [5].
159. Binary with components: white dwarf and red dwarf. Mass of white dwarf = 0.15 $\mathcal{M}_\odot$ mass of red dwarf = 0.14 $\mathcal{M}_\odot$ [61].
164. Mass = 0.94 $\mathcal{M}_\odot$ [50].
165. Member of Praesepe cluster; mass of white dwarf = 0.43 $\mathcal{M}_\odot$ [50].
166. Member of Praesepe cluster; mass = 0.85 $\mathcal{M}_\odot$ [50], $U - V = -0^m.45$ [10].
167. Mass = 0.62 $\mathcal{M}_\odot$ [50].
168. Mass = 0.33 $\mathcal{M}_\odot$ [50]; $W(\mathrm{H}\gamma) = 16$ Å, $W_{0.5}(\mathrm{H}\gamma) = 17$ Å [10].
169. Binary with LDS 235 Å. Component's characteristics: $V_E = 11^m.63$, $V - R = 0^m.64$, $R - I = 0^m.39$, $B - V = 0^m.995$, $U - B = 1^m.00$, $M_v = 6^m.8$. Spectr. type dK3 [10].
173. Binary with LP 90-71. Component's characteristics: $V_E = 15^m.65$, $B - V = 1^m.64$, $U - B = 1^m.18$, $M_v = 12^m.1$. Spectr. type dM2. $\rho = 53''$, $\theta = 63°$ [5]. Spectra is possibly more compose; lines CA in absorption and lines H in emission (DC + dKe) are suspected. Suspected relativistic object. Wide pair. Component-red star is not very late by type.
174. Blue star contains CI atoms and many moleculars C_2. Other known white dwarfs do not contain such peculiarities λ 4670 stars contain only C_2. Proper motion is similar to NB 77408 and stars probably form binary system.
Characteristics: NB 77408: $V_E = 7^m.00$, $B - V = 0^m.49$, $U - B = -0^m.05$, $M_v = 4^m.5$.
Spectr. type dF6.
177. Circular polarization 0.8% $g < 0.5\%$ [5].
Linear $\leq 0.13\%$ [22].
Possibly, variable white dwarf. Mass = 0.77 $\mathcal{M}_\odot$ [50].
178. Binary with G 116-14. Component's characteristics: $V_E = 9^m.00$, $B - V = 0^m.68$, $U - B = 0^m.18$, $M_v = 5^m.4$. Spectr. type dG5, mass = 0.22 $\mathcal{M}_\odot$. $W(\mathrm{H}\gamma) = 19.5$ Å, $W_{0.5}(\mathrm{H}\gamma) = 22$ Å [10].
179. Binary with G 117-b15B with characteristics: $B - V = 1^m.60$, $M_v = 14^m.0$. Spectr. type dM2.
Mass = 0.42 $\mathcal{M}_\odot$ [50]. Variability: $\Delta M = -0^m.44$.
181. Mass = 0.77 $\mathcal{M}_\odot$ [50].
$\Delta M = 0^m.08$ [35].
183. Binary white dwarf with sep. 4″. Spectra of the system is unique, almost continuum [5] $\theta = 36°$. Wide pair. Both components of binary system are dwarfs. Suspected as relativistic object.
185. Mass = 0.26 $\mathcal{M}_\odot$ [50]. $W(\mathrm{H}\gamma) = 39.3$ Å, $W_{0.5}(\mathrm{H}\gamma) = 45$ Å [10].
186. Binary with G 117 B11A: $V_E = 15^m.17$, $V - R = 1^m.28$, $R - I = 1^m.11$, $B - V = 1^m.60$, $M_v = 11^m.2$.
187. Possibly, spectra λ 4670: there is polarization $g < 0{,}20\%$.
189. Mass = 0.74 $\mathcal{M}_\odot$ [50]. $W(\mathrm{H}\gamma) = 15.6$ Å, $W_{0.5}(\mathrm{H}\gamma) = 20$ Å [10].
192. Possibly, star λ 4670; polarization $g < 0.22\%$.
199. There is suspected relativistic object [5].
203. Variable BS dwarf. There are bursts of $0^m.65$ for a min. [5]. Polarization $g < 0.22\%$. Suspected as relativistic object [5].
210. Mass = 0.93 $\mathcal{M}_\odot$ [50].
211. $\Delta V = 0^m.35$; possible variability in brightness, suspected to be relativistic object [5].
212. Mass = 0.46 $\mathcal{M}_\odot$ [50]; $W(\mathrm{H}\gamma) = 32.1$ Å; $W_{0.5}(\mathrm{H}\gamma) = 51$ Å [10].
213. Binary with L970-27; $V_E = 12^m.35$, $V - R = 1^m.29$, $R - I = 1^m.0.05$, $B - V = 1^m.52$, $U - B = 1^m.33$, $M_v = 10^m.8$.
Spectr. type dM6; Mass = 0.55 $\mathcal{M}_\odot$ [50].
217. Spectra may also belong to white dwarf. It may be binary like SS Cygn. High U-velocity indicates that this

star is the star SdO with red component, both of low luminosity. Color is like WZ Sgt. Polarization $g < 0.34\%$ [5].

218. Mass $= 0.57\,\mathcal{M}_\odot$ [50]; $W(H\gamma) = 5.6$ Å, $W_{0.5} = +18$ Å [10].
223. Possibly spectroscopic or unresolved binary system.
228. Possible variability in brightness.
229. Wide binary with G 148-6: $V_E = 14^m.06$, $B - V = 1^m.60$, $U - B = 1^m.17$, $R - I = 1^m.14$, $M_v = 11^m.5$. Spectr. type dM4. Red component has peculiar spectra with weak region TiO. Mass of white dwarf $= 0^m.57\,\mathcal{M}_\odot$ [50]; $W(H\gamma) = 59$ Å, $W_{0.5}(H\gamma) = 39.9$ Å.
230. Binary with L 1405-40 Å: $V_E = 15^m.55$, $V - R = 1^m.26$, $R - I = 1^m.25$, $B - V = 1^m.41$, $U - B = 0^m.73$, $M_v = 12^m.4$.
Spectr. type dM2e. A component has ultraviolet deficiency, which is characteristic of Hyades, but its spectra shows H emission and weak region TiO. Mass of white dwarf $= 0.63\,\mathcal{M}_\odot$ $W(H\gamma) = 31.8$ Å, $W_{0.5}(H\gamma) = 35$ Å.
233. There is polarization $g < 0.20\%$.
239. Binary, component is too close for photometry.
240. Binary with G 148-B4A: $V_E = 15^m.12$, $V - R = 1^m.25$, $R - I = 1^m.15$, $B - V = 1^m.58$, $M_v = 11^m.5$.
246. $W(H\gamma) = 43$ Å, $W_{0.5} = 41$ Å; triplet with G 61-16: $V_E = 13^m.48$, $V - R = 0^m.91$, $R - I = 0^m.75$, $B - V = 1^m.47$, $U - B = 1^m.10$, $M_v = 9^m$. Spectr. type sdM2. Component has weak TiO regions. The third star is at the distance $\rho = 3''$, $\theta = 240°$.
Mass of white dwarf $= 0.79\,\mathcal{M}_\odot$ [50].
249. Spectra is like spectra of hot subdwarf sdO.
252. Possibly weak subdwarf sdO; there is no helium in spectra.
255. Mass of white dwarf $= 0.82\,\mathcal{M}_\odot$ [50].
Luminosity $L = 2 \times 10^{29}$ erg s^{-1}. There is polarization $g < 0.22\%$, $\Delta M = 0^m.74$ [5, 35].
257. By spectra – the late subdwarf with slight lines. Parallax indicates that the star is degenerate [10].
260. Possibly spectroscopic or unresolved binary.
261. Very hot binary. It is suspected to be a relativistic object [5].
263. $W(H\gamma) = 9.2$ Å, $W_{0.5} = 37$ Å. White dwarf mass $= 0.48\,\mathcal{M}_\odot$ [50]. The object has close component, which is possibly variable, photometrically nondistinguishible dM spectra.
267. Binary with W485B: $V_E = 14^m.34$, $V - B = 1^m.82$, $R - I = 1^m.325$, $B - V = 1^m.63$, $U - V = 1^m.63$, $U - B = 1^m.48$, $M_v = 13^m.1$. Spectr. type dM5.
White dwarf mass $= 0.41\,\mathcal{M}_\odot$ [50]. $W(H\gamma) = 31.9$ Å, $W_{0.5}(H\gamma) = 45$ Å, $\Delta M = 0^m.53$ [35].
269. Spectroscopic or unresolved binary is suspected.
271. White dwarf mass $= 0.22\,\mathcal{M}_\odot$ [50]; $W(H\gamma) = 2.2$ Å; $W_{0.5}(H\gamma) = 19$ Å.
272. Binary with LDS 455B: $V_E = 15^m.55$, $B - V = -0^m.83$, $M_v = 11^m.9$; white dwarf mass $= 0.94\,\mathcal{M}_\odot$ [50].
277. White dwarf mass $= 0.68\,\mathcal{M}_\odot$ [50].
280. Binary with G 165-B5A: $V_E = 0^m.09$, $B - V = 0^m.54$, $U - B = -0^m.04$, $M_v = 4^m.90$. Wide pair. Mass of white dwarf $= 0.77\,\mathcal{M}_\odot$.
284. Mass of white dwarf $= 0.58\,\mathcal{M}_\odot$ [50].
299. Mass of white dwarf $= 0.62\,\mathcal{M}_\odot$ [50].
305. Binary with LP 135-154: $V_E = 15^m.74$, $B - V = 1^m.56$, $U - B = 1^m.56$, $M_v = 11^m.0$. White dwarf mass $= 0.22\,\mathcal{M}_\odot$ [50].
313. Binary component: $V_E = 9^m.78$, $B - V = 0^m.53$, $U - B = -0^m.02$, $M_v = 4^m.8$. Spectr. type dG0. White dwarf mass $= 0.83\,\mathcal{M}_\odot$ [50]. Star is very blue for its type.
314. Binary with HR 5864. Component's charac: $V_E = 6^m.02$, $B - V = 0^m.72$, $U - B = 0^m.29$, $M_v = 5^m.5$. Spectr. type dG0. White dwarf mass $= 0.17\,\mathcal{M}_\odot$ [50]. $W(H\gamma) = 29$ Å, $W_{0.5}(H\gamma) = 39$ Å [10].
316. Binary with G 152-B4A: $V_E = 14^m.35$, $B - V = 1^m.56$, $R - I = 1^m.03$, $R = 13^m.03$ [3].
Stars with sep. 10″ have common proper motion 0″12. Spectra, $M_v = +10^m.7$ give $M_v = 11^m.15$ for white dwarf and V_E may be variable for component. $W(H\gamma)$ for white dwarf $= 29$ Å. White dwarf mass $= 0.34\,\mathcal{M}_\odot$ [50].
317. $W(H\gamma) = 30.6$ Å, $W_{0.5}(H\gamma) = 40$ Å [10]. White dwarf mass $= 0.68\,\mathcal{M}_\odot$ [50]. Possibly variable.
319. $W(H\gamma) = 21.8$ Å, $W_{0.5}(H\gamma) = 27$ Å [10].
White dwarf mass $= 1.1\,\mathcal{M}_\odot$ [50].
327. White dwarf mass $= 0.94\,\mathcal{M}_\odot$ [50].
332. Binary with G 225-67. Component's charac: $B - V = 1^m.60$, $U - B = 1^m.05$, $V = 12^m.90$. Spectra dM3-dM4. Sep. $\rho = 25''$, $\theta = 18°$. Component – red dwarf CM Dra – flare star. In general, this object is an eclipsing system with a period $P = 0^d.63398$ and with depth of minimum in V filter $= 0^m.7$.

$\Delta V = V - m_{1.85} = 0^m.7$. By [35] spectr. type of component will determine like dM5e. White dwarf mass $= 0.62\,\mathcal{M}_\odot$ [50]; luminosity $L = 7.5 \times 10^{29}$ erg s^{-1} [5]. By [52] white dwarf mass $= 0.252\,\mathcal{M}_\odot$ red dwarf mass $= 0.235\,\mathcal{M}_\odot$. By [49] white dwarf mass $= 0.24\,\mathcal{M}_\odot$, red dwarf mass $= 0.21\,\mathcal{M}_\odot$.

336. $W(H\gamma) = 25.3$ Å, $W_{0.5}(H\gamma) = 26$ Å; white dwarf mass $= 0.52\,\mathcal{M}_\odot$ [50].
342. Possibly subdwarf sdO.
343. White dwarf mass $= 0.57\,\mathcal{M}_\odot$ [50].
344. White dwarf mass $= 0.48\,\mathcal{M}_\odot$ [50]. Possibly variable.
345. White dwarf mass $= 0.32\,\mathcal{M}_\odot$ [50]. Probably binary with HR 6314.
346. White dwarf mass $= 0.74\,\mathcal{M}_\odot$ [50].
347. Wide binary with 181-B5A = 33°2834. Component's characteristics: $V_E = 8^m.68$, $B - V = 0^m.545$, $U - B = 0^m.00$, $M_v = 4^m.8$. Mass of white dwarf $= 0.76\,\mathcal{M}_\odot$ [50].
348. There is polarization $g < 0.14\%$.
355. Binary with W 672 A. Component's characteristics: $V_E = 14^m.00$, $V - R = 1^m.43$, $R - I = 1^m.21$, $B - V = 1^m.54$, $U - B = 1^m.33$, $M_v = 11^m.9$. Spectral characteristics of component Sd M6. Mass of white dwarf $= 0.38\,\mathcal{M}_\odot$ [50]. $W(H\gamma) = 37$ Å, $W_{0.5}(H\gamma) = 48$ Å [10].
359. Mass of white dwarf $= 0.51\,\mathcal{M}_\odot$ [50]. $\Delta M = -0^m.43$ [35].
361. Binary with G 154-B5A. Component's characteristics: $V_E = 11^m.91$, $B - V = 1^m.44$, $U - B = 1^m.20$, $R - I = 0^m.735$, $M_v = 8^m.9$. Spectr. type dM3. Sep. components 33″ and common proper motion 0″.99. $M_v = 8^m.9$ gives for red component. Mass of white dwarf $= 0.21\,\mathcal{M}_\odot$ [50].
362. Magnetic white dwarf. Field $= 2$–4×10^7 G [60].
363. Polar white dwarf recently discovered.
Circular polarization 0.5%, linear 1.5%. Mass of white dwarf $= 0.78\,\mathcal{M}_\odot$ [50].
364. Binary with G 140-B1A. Component's characteristics: $V_E = 9^m.36$, $V - R = 0^m.46$, $R - I = 0^m.33$, $B - V = 0^m.95$, $U - B = 0^m.795$, $M_v = 6^m.3$.
Spectr. type of component dK2. Sep. $\rho = 27''$, $\theta = 204°$.
There is polarization of white dwarf $g < 0.22\%$; luminosity 5×10^{30} erg s^{-1} [5].
365. This star LP 9-231 is so-called ‘pygmy’. Mass of white dwarf $= 0.48\,\mathcal{M}_\odot$ [50].
367. Binary white dwarf (35). Binary with G 206-17. Component's characteristics: $m_{pg} = 16^m.2$. Spectr. type DA_{wk}.
368. Separation $\rho = 55''$, $\theta = 76°$.
370. Mass of white dwarf $= 0.15\,\mathcal{M}_\odot$ [50]; $B - m_{pg} = 0^m.7$ [5].
374. Mass of white dwarf $= 0.45\,\mathcal{M}_\odot$ [50]; $W(H\gamma) = 14.7$ Å, $W_{0.5}(H\gamma) = 23$ Å [10].
376. Recently discovered polar. white dwarf.
There is weak absorption. Linear polarization 0.1%, circular 3%.
381. White dwarf mass $= 0.55\,\mathcal{M}_\odot$ [50]; $W(H\gamma) = 47.9$ Å; $W_{0.5} = 51$ Å [10].
384. Magnetic field magnitude 50 (MG)? Circular polarization 3%, linear 2% [35].
385. Binary with G 142-B2A. Component's characteristics: $V_E = 12^m.68$, $V - R = 1^m.16$, $R - I = 0^m.95$, $B - V = 1^m.54$, $U - B = 1^m.3$, $M_v = 10^m.2$. Spectr. type Sd M2. By (39) supposed initial mass of the star resulting in the white dwarf was 4-6$\mathcal{M}_\odot$. Final white dwarf mass $\approx 0.73\mathcal{M}_\odot$. Member of the Sirius group. By [50] final white dwarf mass $= 0.38\,\mathcal{M}_\odot$.
386. White dwarf mass $= 0.83\,\mathcal{M}_\odot$ [50]. Luminosity $L = 4 \times 10^{29}$ erg s^{-1}.
387. Binary with LDS 678 A. Component's characteristics: $V_E = 12^m.12$, $12^m.24$, $V - R = 1^m.40$, $R - I = 1^m.09$, $B - V = 1^m.63$, $U - B = 1^m.42$, $M_v = 12^m.5$.
Spectr. type dM5. Sep. of components $\rho = 27''2$, $\theta = 308°$. White dwarf mass $= 0.71\,\mathcal{M}_\odot$ [50]; luminosity $L = 5 \times 10^{30}$ erg s^{-1}.
391. Binary with LDS 683 A. Component's characteristics: $V_E = 13^m.60$, $V - R = 0^m.98$, $R - I = 0^m.725$, $B - V = 1^m.49$, $U - B = 1^m.4$, $M_v = 8^m.8$. Spectr. type Sd MI. White dwarf mass $= 0.56\,\mathcal{M}_\odot$ [50].
392. White dwarf mass $= 0.37\,\mathcal{M}_\odot$ [50].
395. White dwarf mass $= 0.66\,\mathcal{M}_\odot$ [50]. $W(H\gamma) = 28.4$ Å, $W_{0.5} = 44$ Å [10].
399. White dwarf mass $= 0.68\,\mathcal{M}_\odot$ [50]; $W(H\gamma) = 12$ Å, $W_{0.5} = 21$ Å [10]. $U - V = -0^m.35$ [4].
401. Nova Sge (1913, 1946) with period 0^d.0567. White dwarf mass $= 0.67\,\mathcal{M}_\odot$, red dwarf mass 0.01 $\mathcal{M}_\odot$. By [39] mass of star progenitor mass $\lesssim 8\,\mathcal{M}_\odot$.
405. There is polarization: circular $\sim 1\%$, linear $\sim 3\%$ [22].
406. Binary with G 24-10. Component's charac: $V_E = 13^m.20$, $V - R = 1^m.40$, $R - I = 1^m.20$, $B - V = 1^m.54$, $U - B = 1^m.15$, $M_v = 11^m.8$. Spectr. type dM5. Separation of components $\rho = 102''$, $\theta = 331°$. White dwarf mass $= 0.78\,\mathcal{M}_\odot$ [50]; there is polarization $g < 0.18\%$, luminosity $L = 2 \times 10^{30}$ erg s^{-1} [35].

410. White dwarf mass = 0.59 $\mathcal{M}_{\odot}$ [50]; $\Delta M = -0^m.07$.
417. Very faint photograph. Magnitude $m_{pg} = 17^m.8$. Possibly, it is not a star of the Main Sequence, but degenerate red star, most probably very red BS-dwarf.
Wide binary with R 193A. Component's characteristics: $V_E = 11^m.87$, $B - V = 1^m.49$, $U - B = 1^m.06$, $M_v = 10^m.0$. Spectr. type dM4e.
Sep. system 13″8, $\theta = 317°$. White dwarf mass = 0.22 $\mathcal{M}_{\odot}$ [50].
420. Luminosity $L = 4 \times 10^{30}$ erg s^{-1}. There is polarization $g < 0.20\%$. $\Delta M = -0^m.17$ (variability in brightness). Suspected to be a relativistic object in [5].
421. White dwarf mass = 0.44 $\mathcal{M}_{\odot}$.
426. White dwarf mass = 0.40 $\mathcal{M}_{\odot}$ [50].
430. White dwarf mass = 0.84 $\mathcal{M}_{\odot}$ [50].
431. White dwarf mass = 0.35 $\mathcal{M}_{\odot}$ [50].
432. Binary with LDS 749A. Component's characteristics: $V_E = 9^m.89$, $V - R = 0^m.60$, $R - I = 0^m.47$, $B - V = 0^m.96$, $U - B = 0^m.48$, $M_v = 7^m.2$, spectr. type Sd K4. White dwarf mass = 0.4 $\mathcal{M}_{\odot}$ [50].
435. White dwarf mass = 0.32 $\mathcal{M}_{\odot}$ [50].
438. White dwarf mass = 0.44 $\mathcal{M}_{\odot}$ [50]; $\Delta m_{pg} = 1^m.4$, probably, variability in brightness [5]. Luminosity $L = 5 \times 10^{30}$ erg s^{-1}.
441. White dwarf mass = 0.35–0.45 $\mathcal{M}_{\odot}$ [50].
444. White dwarf mass m = 0.52 $\mathcal{M}_{\odot}$ [50].
447. Common proper motion with component of late type GD-51°13/28.
448. Binary? Component of M-type at the distance 19″.
450. Spectr. type BS. However, There are Balmer lines, but they are rather rough and wide. White dwarf mass = 0.87 $\mathcal{M}_{\odot}$ [50].
458. White dwarf mass = 0.93 $\mathcal{M}_{\odot}$ [50].
459. White dwarf mass = 0.31 $\mathcal{M}_{\odot}$ [50].
461. Binary system. Component's charac: $V_E = 8^m.00$, $B - V = 0^m.65$, $U - B = 0^m.06$, $M_v = 4^m.7$.
Spectr. type G6, component 178A = HD 216777 – bright star with radial velocity – 23 km s^{-1}. Sep. system 43″ and common proper motion 0.59″. By (39) supposed initial mass of star – white dwarf ≈ 1–1.1 $\mathcal{M}_{\odot}$, but final ≈ 0.25 $\mathcal{M}_{\odot}$. By [50] white dwarf mass = 0.45 $\mathcal{M}_{\odot}$.
463. Possibly composition DK + DA. Possible variability. Possibly member of group W 219 [3].
469. White dwarf mass = 0.67 $\mathcal{M}_{\odot}$ [50].
470. Possibly peculiar hot subdwarf.
472. $W(H\gamma) = 4.7$ Å, $W_{0.5}(H\gamma) = 10$ Å.
Evidently a peculiar hot subdwarf.
476. White dwarf mass = 0.84 $\mathcal{M}_{\odot}$ [50]. Possibly old star.
478. White dwarf mass = 0.57 $\mathcal{M}_{\odot}$ [50].
481. Binary with L 1512 = 34A. Component's characteristics: $V_E = 11^m.68$, $V - R = 1^m.25$, $R - I = 1^m.10$, $B - V = 1^m.55$, $U - B = 1^m.12$, $M_v = 11^m.2$, spectr. type dM5. White dwarf mass = 0.49 $\mathcal{M}_{\odot}$ [50].
485. Binary? Component of the late type is at the distance 7″.
487. White dwarf mass = 1.31 $\mathcal{M}_{\odot}$ [50].

White Dwarfs

No.	Name of star	α_{1950}	δ_{1950}	Sp. type	$T_{\rm eff}$ or θ_e	M_v	$\log R/R_0$	Referen
(1)	(2)	(3)	(4)	(5)	(6)	(7)	(8)	(9)
		0^h00^m2	$-34°30'$	DC?	6700 K			5
2	Gr 305							
	GD 408	$00^h02^m28^s$	$+72°56'4$	DB		11^m22		35
3	EG 203							
	GD 2	$00^h04^m58^s$	$+33°00'8$	DA_{wk}	$\log\theta = 0.17$		2.02; 2.10	4
4	EG 1	00^h07^m0	$+30°52'$	DC	$\log\theta = 0.11$	13^m4		10
	G 130-49					13^m18 ptm		
5	Gr 381	$00^h09^m39^s$	$+50°09'0$	DC?				2
	HS			DA–F		13^m94 ptm		35
	G 217-37							
6	EG 2	00^h11^m1	$+0°03'$	DA		11^m3		10
						13^m0		
	W 1				9.4×10^3 K		$R = 0.0104\,R_0$	50
	G 31-35							
7	EG 3	00^h17^m4	$+13°36'$	DB		10^m8		10
	LB 433							
	F 4							
8	Gr 306	$00^h30^m19^s$	$44°27'7$	DA		12^m73		35
	G 172-4				9.6×10^3 K		$R = 0.0172\,R_0$	50
9	Gr 337							
	GD 411							
	LP 12-438	$0^h33^m13^s$	$+77°07'$	DA				36
10	Gr 382							
	WH	$00^h31^m22^s$	$-12°24'4$	DC?		14^m66		35
	158-100			DK–G				
11	EG 4				11.2×10^3 K		$R = 0.0081\,R_0$	50
	L 1011-71	00^h33^m1	$+1°37'$	DA		$+11^m3$		
	G 1-7					$+12^m8$		10
					0.55			
				DA		12^m32		35
12	EG 204	$00^h37^m15^s$	$+31°16'1$	DA_{wk}	$\log\theta = 0.18$		1.90; 2.18	4
	GD 8							
13	Gr 383				$\log\theta' = 0.10$			
	G 242-63	$0^h38^m31^s$	$73°4'8$	DB		12^m12		35
14	EG 245							
	GD 218-8	00^h38^m5	$+55°33'7$	DC				5
	LTT 17144							
15	EG 246							
	LTT 375	00^h39^m0	$-22°38'$	DC	6000 K			5
	L 651-57				5600 K			
16	Gr 267							
	F 7	00^h41^m2	$-10°17'$	DC?				5
	G 270-48			DC		11^m43		35
				H, He		11^m48		22
17	EG 5							
	V Maan 2	00^h46^m5	$+5°10'$	DG		$+14^m2$		10
18	G 1-27					14^m8		47
	G 21-27							9
				DG		13^m90		35
	BPM 16274	00^h47^m8	$-52°25'$	DA–F				58

ntinued)

Name of star	α_{1950}	δ_{1950}	Sp. type	T_{eff} or θ_e	M_v	log R/R_0	References
(2)	(3)	(4)	(5)	(6)	(7)	(8)	(9)
Anoun	00^h50^m9	−33°16′	DA				58
G 69-31	$00^h52^m05^s$	+22°40′4	DA				2
EG 6							
L 796-10	00^h53^m3	−11°45′	(DA)		$+11^m5$		10
L 7					$+13^m4$		
GD 9	$00^h58^m24^s$	−04°27′5	DA				1
EG 7				8.6×10^3 K		$R = 0.0151\ R_0$	50
G 1-45	1^h01^m2	+4°42′	DA_s		$+11^m4$		10
					$+12^m8$		
						−1.71	9
				0.52		−1.99	
			DA				35
					13^m20		
Gr 338							
GD 273	$1^h03^m19^s$	+55°49′	DA		15^m5		36
T 191	1^h04^m4	−33°09′	DA				58
BPM 30551	1^h04^m7	−46°26′	DA				58
EG 179							
GD 11	1^h06^m6	+37°17′	DA		10^m6		10
			DAs				38
EG 8							
L 1373-25	1^h07^m5	+26°45′	DA		11^m2		10
LTT 10425					12^m7		
EG 9							
W 1516	1^h15^m4	+15°56′	DC		11^m0		10
L 8					12^m6		
				10000 K	12^m7		5
				10600 K			
Gr 307							
LTT 784	1^h23^m0	−26°14′	DC?		13^m58		35, 5
G 269-160							
EG 205	$1^h26^m40^s$	+42°12′8	DA				4
GD 13							
G 2-40	$01^h26^m46^s$	10°07′9	DAk				1
				9.2×10^3 K		$R = 0.0189\ R_0$	50
GD 14	$01^h27^m19^s$	+27°00′9	DA				1
				10×10^3 K		$R = 0.0318\ R_0$	50
EG 10							
R 548	1^h33^m7	−11°36′	(DA)		11^m3		10
L 9					13^m0		
PHL 1062							
EG 308							
GD 419	$1^h34^m55^s$	83°19′8	DA		10^m24		35
EG 11				7.3×10^3 K		$R = 0.0197\ R_0$	50
G 271-115	1^h35^m4	−5°14′	DAs		11^m8		10
L 870-2			DA		11^m76		35
LP 648-24					13^m60 ptm		
L 10			DAs		12^m46		59
Gr 339							
GD 420	$1^h36^m45^s$	76°54′	DA		15^m		36
LP 13-249							

(continued)

No.	Name of star	α_{1950}	δ_{1950}	Sp. type	T_{eff} or θ_e	M_v	log R/R_0	Reference
(1)	(2)	(3)	(4)	(5)	(6)	(7)	(8)	(9)
38	EG 12							
	G 34-49	1^h42^m6	+23°09′	DC		$+13^m.7$		
					10 500 K			5
39	EG 13							
	G 94-9	1^h43^m9	+21°40′	DA		$11^m.3$		10
						$12^m.8$		
40	Gr 309							
	PHL 3802	1^h46^m4	−26°52′	DC				5
				Sd B		$11^m.88$ ptm		35
41	Gr 310							
	GD 241	$1^h47^m22^s$	67°24′6	DA		$9^m.80$ ptm		35
42	EG 269	$1^h48^m56'$	+46°45′2					4
	GD 279							
43	EG 14							
	G 71-B5B	1^h50^m5	+8°57′			$12^m.8$		10
44	EG 311							
	G 71-41	$1^h51^m38^s$	1°46′7	DA		$11^m.57$ ptm		35
45	EG 206	$1^h55^m04^s$	+6°57′5	DA				4
	GD 20							
	F 17							
46	EG 15				20.3×10^3 K		$R = 0.0133\, R_0$	50
	Ox 25°6725	2^h05^m9	+25°00′	(DA)		$+10^m.8$		10
	L 11				0.36		1.69	4
					0.24			
47	EG 168			DAs	7.4×10^3 K		$R = 0.0128\, R_0$	50
	G 74-7	$2^h08^m13^s$	39°41′5	DA–F		$13^m.50$		35
						$13^m.62$ ptm		
48	Gr 312							
	GD 25	$02^h13^m13^s$	39°37′6	DA		$12^m.87$ ptm		35
49	EG 16				5.5×10^3 K		$R = 0.0109\, R_0$	50
	G 134-22	2^h13^m8	+42°44′	DC		$14^m.8$		10
					5000 K			5
				DC		$14^m.61$ ptm		35
50	EG 17							
	h Per 1166	2^h14^m0	56°53′	(DA)	$\log \theta_e = 0.28$	$+10^m.7$		10
	L 12						1.80	4
					$\log \theta_e' = 0.19$		2.08	
51	EG 18				16.3×10^3 K		$R = 0.0068\, R_0$	50
	G 94-B5B	2^h20^m7	22°14′	DAn		$12^m.5$		10
52	EG 19							
	F 22	2^h27^m7	5°03′	DA		$+10^m.9$		10
					0.23		1.78	9
53	GD 30	$2^h30^m.37$	34°20′2	DA				1
54	EG 207							
	GD 31	$2^h31^m37^s$	−5°24′9		$\log \theta = 0.53$		1.60; 1.96	
	PHL 1358				$\log \theta' = 0.36$		1.88; 2.24	
55	EG 20							
	F 24	2^h32^m5	3°31′	DAwk		$10^m.4$		10
				DA k		$10^m.52$ ptm		35
					60×10^3 K		$R = 0.0239\, R_0$	50
56	EG 314							
	G 174-5	$2^h32^m50^s$	52°31′2	DA		$10^m.98$ ptm		35

tinued)

Name of star	α_{1950}	δ_{1950}	Sp. type	T_{eff} or θ_e	M_v	log R/R_0	References
(2)	(3)	(4)	(5)	(6)	(7)	(8)	(9)
HS							
G 4-34	$2^h39^m26^s$	10°59′9	DC				2
G 174-14	$2^h45^m03^s$	54°11′1	DF wk				2
LFT 245	2^h55^m8	−70°34′	DA				58
				7.3×10^3 K		$R = 0.0111\ R_0$	50
G 176-48	$2^h57^m17^s$	8°00′0	DA wk				2
Gr 287							
GD 38	$2^h59^m20^s$	37°49′3	DA				1
EG 384							
GD 40	$3^h00^m21^s$	−1°20′2	DB				1
			DB		$11\overset{m}{.}57$ ptm		35
BPM 17088	3^h08^m5	−56°34′	DB				58
EG 21[a]							
LB 3303	3^h10^m0	−68°48′	DA		$+10\overset{m}{.}0$		10
EG 21							
G 5-28	$3^h15\overset{m}{.}7$	15°02′	DMp?		$14\overset{m}{.}1$		10
			Sd M				35
EG 208							
GD 45	$3^h16^m34^s$	+34°31′6	DAs	log $\theta = 0.47$		1.64; 2.06	
BPM 85584				log $\theta' = 0.30$		1.92; 2.30	4, 38
G 77-50	$3^h22^m39^s$	−01°59′2	DG				2
EG 385							
HS	$3^h24^m35^s$	73°51′7	DC				2
G 221-10			DF–C		$13\overset{m}{.}37$		35
EG 22							
L 587-77A	3^h26^m8	−27°34′	DAs				10, 5
L 14					$11\overset{m}{.}$		
EG 23							
G 37-44	3^h32^m1	32°02′	DA		$11\overset{m}{.}3$		10
					$13\overset{m}{.}0$		
				0.54		1.68	9
GD 47	$3^h39^m24^s$	−03°32′3	DA				1
EG 24							
G 6-30	3^h41^m6	18°18′	4670		$14\overset{m}{.}4$		10
W 219					$14\overset{m}{.}58$		35
L 15					$14\overset{m}{.}4$		
					$13\overset{m}{.}81$ ptm		
BPM 31594	3^h41^m8	−45°58′	DA				44
BPM 3116	3^h42^m0	−67°19′	DA				44
GD 52	3^h48^m48	33°58′6	DA				1
EG 25				29.8×10^3 K		$R = 0.0081\ R_0$	50
LB 1497	3^h49^m1	24°47′	DAwk		$+11\overset{m}{.}0$		10
					$11\overset{m}{.}0$		35
			DA		$10\overset{m}{.}41$ ptm		39
				15.1×10^3 K		$R = 0.0120\ R_0$	50
EG 26							
HG 7-41	3^h52^m1	9°37′	DA		$11\overset{m}{.}5$		10
HZ 4				0.52		1.73	9
			DA		$11\overset{m}{.}20$		35
EG 27							
G 7-17	3^h58^m7	18°36′	DM		$14\overset{m}{.}7$		10
			Sd M				35

(continued)

No.	Name of star	α_{1950}	δ_{1950}	Sp. type	T_{eff} or θ_e	M_v	log R/R_0	References
(1)	(2)	(3)	(4)	(5)	(6)	(7)	(8)	(9)
79	EG 28				13.5×10^3 K		$R = 0.0109\,^R$	50
	H–G 8-7	4^h01^m5	25°01′	DA		$12^m.3$		10
	LB 1240						1.89	9
	G 8-8				0.51			
	LP 357-134			DA		$11^m.69$		35
						$11^m.69$ ptm		
80	EG 29				15.3×10^3 K		$R = 0.0085\,R_0$	50
	HG 7-112	4^h06^m3	16°59′	DA		$12^m.4$		10
	LB 227			DA		$12^m.06$		35
	LP 414-10					$11^m.45$ ptm.		
81	EG 30							
	HZ 10	4^h07^m3	17°54′	DA		$+11^m.1$		10
	L 17				0.52		1.70	9
82	GD 56	$+04^h08^m33^s$	−4°06′1	DA				1
83	EG 31							
	HZ 2	4^h10^m2	11°45′	DA		$+10^m.9$		10
					0.36		1.75	9
				DA		$10^m.57$ ptm		35
84	EG 32							
	HG 7-138	4^h12^m4	14°57′	DK		$12^m.6$		10
85	EG 33				16.9×10^3 K		$R = 0.0124\,R_0$	50
	GL 166B	4^h13^m0	−7°44′	DA		$+11^m.0$		10
	40 Eri B			DA		$11^m.07$		10
						$11^m.24$ ptm		
86	GD 60	$4^h16^m57^s$	33°28′5	DA				1
87	EG 34							
	G 38-29	4^h17^m0	36°09′	DA s		$11^m.3$		10
						$13^m.0$		
88	EG 35							
	LB 212	4^h18^m9	15°22′	DA wk		$11^m.2$		10
						$12^m.7$		
89	BPM 17731	4^h18^m1	−53°58′	DB				44
90	EG 36				19.4×10^3 K		$R = 0.0113\,R_0$	50
	VR 7	4^h21^m0	16°14′	(DA)		$11^m.3$		10
	HG 7-191							
91	EG 169							
	LB 1320	$4^h23^m07^s$	12°05′	DC?				5
	LP 475-70							
	G 83-10							
92	EG 37				23.9×10^3 K		$R = 0.0110\,R_0$	50
	VR 16	4^h25^m7	16°52′	(DA)		$11^m.0$		10
	HG 7-233							
93	EG 180							
	G 175-346	4^h26^m8	58°53′	DC		$13^m.8$		3
	2051 (St)			DAF–C	7500_K	$13^m.7$		5
						$13^m.66$		35
94	EG 38							
	HG 7-255	4^h29^m4	17°38′	DAe		$11^m.0$		10
	HZ 9							5
				M + DA				35
95	EG 39				20.9×10^3K		$R = 0.0108\,R_0$	50
	HZ 7	4^h31^m0	12°35′	DA		$11^m.2$		10
					0.37		1.83	9
				DA		$11^m.03$		35
						$10^m.52$		

ntinued)

Name of star	α_{1950}	δ_{1950}	Sp. type	T_{eff} or θ_e	M_v	log R/R_0	References
(2)	(3)	(4)	(5)	(6)	(7)	(8)	(9)
EG 40				5.7×10^3 K		$R = 0.0106\ R_0$	50
G 39-27	4^h33^m6	27°04′	DC		$14^m.9$		10
				5400 K	$14^m.67$		59
Gr 315							
GD 61	$4^h35^m10^s$	41°03′7	DBs				1
					$10^m.90$		35
EG 41							
L 879-14	4^h35^m4	−8°53′	4670		$+11^m.2$		10
					$+12^m.7$		
EG 316							
LP 475242	$4^h37^m36^s$	13°53′0	DB		$11^m.64$		35
					$11^m.41$		
EG 42				28.8×10^3 K		$R = 0.0098\ R_0$	50
HZ 14	4^h38^m2	+10°52′	DAn		$10^m.8$		10
				0.25		1.89	9
			DA		$10^m.66$		35
					$10^m.03$ ptm		39
Gr 317							
G 175-46	$4^h40^m05^s$	51°00′8	DA				2
					$13^m.04$ ptm		35
LP 476-267							
G 83-43	4^h52^m09	10°23′3	DAs				2
EG 209	$4^h53^m50^s$	41°51′5	DA	log $\theta = 0.53$		1.66; 2.02	4, 38
GD 64				log $\theta' = 0.43$		1.84; 2.20	
G 191-16	$4^h55^m21^s$	55°20′8	DA		$11^m.78$		35
EG 247			DA wk	61.9×10^3 K		$R = 0.0164\ R_0$	50
G 191-B2B	$5^h01^m30^s$	52°46′0	DA		$8^m.36$		35
					$9^m.56$		
				log $\theta = 0.13$			4
				0.09		(2.08)	
GD 66	$5^h17^m25^s$	30°45′5	DA				
EG 43				8.6×10^3 K		$R = 0.0134\ R_0$	50
G 86-B1B	5^h18^m4	33°19′	DAe		$12^m.7$		10
Gr 319							
GD 69	$5^h32^m48^s$	41°28′2	DAs				1
					$13^m.43$		35
EG 340							
G 98-18	$5^h33^m56^s$	32°13′5	DA		$0^m.0$		35
					$11^m.76$ ptm		
EG 341							
GD 290	5^h43^m35	57°58′3	DA		$12^m.77$ ptm		35
EG 248							
G 99-37	$5^h48^m46^s$	−0°11′2	DC 2p		$14^m.23$ ptm		35
			DGp				
			CH, C_2		$14^m.28$		22
EG 210	$5^h49^m34^s$	+15°52′7	DA wk	log $\theta = 0.17$		2.00	4
LTT 11733				log $\theta' = 0.10$		2.22	
GD 71							
L 1243-36							
EG 44							
G 102-39	5^h51^m1	+12°24′	DC		$12^m.9$		10
				10600 K	$11^m.6$		5
				12000 K			

(continued)

No.	Name of star	α_{1950}	δ_{1950}	Sp. type	T_{eff} or θ_e	M_v	log R/R_0	References
(1)	(2)	(3)	(4)	(5)	(6)	(7)	(8)	(9)
114	EG 45				4.4×10^3 K		$R = 0.0124\,R_0$	50
	LP 658-2	5^h52^m7	−4°09′	DKe		$15^m.4$		10
	G 99-44			DK				35
	HL 4					$15^m.21$ ptm		
115					5.7×10^3 K		$R = 0.0115\,R_0$	50
	Gr 290	5^h53^m8	5°22′	DC	5700 K			5
	G 99-47			DCp		$14^m.55$		35
						$14^m.53$		
				H		$14^m.58$		22
116	EG 320							
	LTT 2437	$5^h59^m24^s$	−12°30′	RSL		$14^m.62$		35
117	G 105-4	$5^h59^m37^s$	15°15′3	DA wk				2
118	LP 33-276	$6^h01^m06^s$	73°32′	(DC)		$(+14^m)$		
119	EG 211	$6^h06^m51^s$	+28°15′1	DA	$\log\theta = 0.36$		1.98; 2.16	4, 38
	GD 72				$\log\theta' = 0.26$		1.98; 2.36	
120	EG 46							
	L 1244-26	6^h12^m4	17°45′	DA		$10^m.7$		10
	G 104-27				26.7×10^3 K		$R = 0.0109\,R_0$	50
121	L 182-61	6^h15^m5	−59°11′	DB		$12^m.1$		58
122	HS	$6^h18^m05^s$	6°46′7	DC				2
	G 105-30							
123	BPM 4225	6^h24^m4	−75°40′	DA				44
124	EG 47							
	G 105-B2B	6^h25^m0	10°02′	DF		$13^m.2$		10
125	GD 74	$6^h25^m30^s$	41°32′8	DA				1
126	G 107-9	$6^h32^m59^s$	40°57′1	DA				2
127	EG 212	$6^h37^m26^s$	+47°47′3	DA	$\log\theta = 0.50$		1.66; 2.02	
	GD 77				$\log\theta' = 0.33$		1.92; 2.28	
128	EG 48							
	SA 26-82	6^h39^m6	44°43′	(DA)		$10^m.8$		10
	L 22							
129	Sirius							
	EG 49				29×10^3 K		$R = 0.0074\,R_0$	50
	αCMa B	6^h43^m0	−16°39′	DA		$11^m.1$		10
130	EG 50				22.2×10^3 K		$R = 0.0094\,R_0$	50
	G 87-7	6^h44^m3	37°36′	DAn		$+11^m.0$		10
	He 3			DA		$10^m.81$		35
	L 23					$10^m.35$ ptm		
131	G 108-26	$6^h44^m47^s$	2°34′4	DAs				2
132	EG 213	$6^h51^m42^s$	−2°04′6	DAwk				5
	GD 80							
133	EG 342							
	G 250-26	$6^h48^m40^s$	64°7′6	DA–F		$14^m.38$		35
	LP 58-53					$14^m.5$		36
134	GD 78	$6^h48^m51^s$	36°49′4	DC	5500 K			5
	WH							
135	EG 181							
	G 108-42	6^h54^m8	2°45′	DC		$11^m.2$		3, 10
						$12^m.8$		
					9500 K			5
136	BPM 18394	7^h01^m5	−58°46′	DA				44

ntinued)

.	Name of star	α_{1950}	δ_{1950}	Sp. type	T_{eff} or θ_e	M_v	log R/R_0	References
	(2)	(3)	(4)	(5)	(6)	(7)	(8)	(9)
	EG 51							
	G 87-29	7^h06^m9	37°45′			$13^m.4$		10
	EG 214	$7^h10^m24^s$	+21°39′2	DAs				38
	GD 83							
	GD 294	7^h13^m20	58°29′8	DA				1
	EG 215							
	GD 84	$7^h14^m23^s$	45°53′4	DC?	10 800			5
				DC				38
	EG 216	$7^h16^m32^s$	+40°27′1	DB	log θ = 0.28		1.78	4
	GD 85			DBs	log θ' = 0.17		2.10	38
	EG 52				4.3×10^3 K		$R = 0.0149\ R_0$	50
	G 107-70	7^h27^m1	48°17′4	DC		$14^m.6$		10
	G 193-51			DC	4100 K			5
	G 111-8							
						$15^m.09$		59
				DC		$14^m.70$		35
						$15^m.14$ ptm		
	Gr 321							
	G 234-4	7^h28^m8	64°16′	DK–DC				5
				DC–K		$14^m.86$ ptm		35
	GD 86	$7^h30^m44^s$	48°47′9	DA				1
	L 384-24	7^h32^m1	−42°47′	DAc				58
	EG 53							
	αCMiB	7^h36^m7	5°21′			$13^m.0$		10
	Procion							
	EG 54							
	L 745-46A	7^h38^m0	−17°17′	DF		$13^m.5$		10
	L 25				0.60		2.06	9
				DF		$13^m.67$		35
						$13^m.33$ ptm		
	BPM 18615	7^h40^m8	−57°02′	DF				58
	GD 343							
	GD 89	$7^h43^m54^s$	44°16′3	DAn				1
				DA		$12^m.21$ ptm		35
	Gr 344							
	G 193-74	$7^h49^m37^s$	52°37′	DC				5
						$13^m.44$		35
	EG 55							
	NGC 2477-116	7^h49^m8	−38°21′			$10^m.7$		10
	Gr 322							
	G 193-78	7^h51^m9	57°50′	DC				2
				DC		$12^m.51$		35
	Gr 345				7.9×10^3 K		$R = 0.0103\ R_0$	50
	G 90-28	$7^h52^m09^s$	36°31′1	DA				2
				DA–C		$13^m.40$		35
				DAwk		$13^m.34$ ptm		
	EG 56							
	97-12	7^h52^m8	−67°38′	DC		$13^m.0$		10
				DC?	5700 K	$15^m.3$		5
	EG 57							
	L 817-13							10
		7^h52^m8	−14°38′	DA		$10^m.9$		10
						$12^m.2$		
					0.39		1.74	
							2.00	9

(continued)

No.	Name of star	α_{1950}	δ_{1950}	Sp. type	T_{eff} or θ_e	M_v	log R/R_0	References
(1)	(2)	(3)	(4)	(5)	(6)	(7)	(8)	(9)
156	BPM 18764	8^h00^m7	$-53°19'$	DAe				44
157	Gr 346							
	G 111-54	$8^h02^m18^s$	$38°41'0$	DF		$12^m.38$		35
				DFp		$12^m.14$ ptm		
158	L 97-3	8^h06^m4	$-66°09'$	DC?				44
159	Z Cha	$8^h09^m56^s$	$-76°14'6$	DA				55
								56
160	Gr 347							
	GD 295	$8^h10^m44^s$	$23°20'7$	DA		$11^m.40$		35
161	Gr 368							
	GD 90	$8^h16^m31^s$	$37°40'9$	WDpec		$12^m.63$		1
				DAp		$12^m.58$ ptm		35
162	EG 58				8.0×10^3 K		$R = 0.0100\, R_0$	50
	G 111-71	8^h16^m7	$38°44'$	DA, F?		$13^m.2$		10
					0.62		1.97	9
163	EG 217	$8^h26^m42^s$	$+45°30'6$	DAs	$\log\theta = 0.53$		1.74, 2.06	4
	W 309				$\log\theta' = 0.45$		1.86, 2.22	
	GD 91							
	LTT 12215							
164	Gr 249	8^h27^m	$+32°$	DAs	8.1×10^3 K		$R = 0.0084\, R_0$	50
165	EG 59				10.6×10^3 K		$R = 0.0141\, R_0$	50
	LB 390	8^h36^m9	$20°11'$	DA		$12^m.2$		10
166	EG 60				13.3×10^3 K		$R = 0.0093\, R_0$	50
	LB 1847	8^h36^m9	$19°57'$	DA		$12^m.6$		10
167	EG 61			DA	13.6×10^3 K		$R = 0.0118\, R_0$	50
	LB 393	8^h37^m6	$19°54'$			$12^m.1$		10
168	EG 62				9.3×10^3 K		$R = 0.0156\, R_0$	50
	LP 532-81	8^h39^m7	$-32°47'$	DA, Fs		$12^m.0$		10
	L 28							9
169	EG 63							
	LDS 235B	8^h45^m3	$-18°48'$	DB		$10^m.7$		10
	L 29					$11^m.44$		35
170	Gr 329							
	GD 96	8^h46^m2	$34°40'9$	DA		$13^m.45$		35
171	BPM 5109	8^h50^m0	$-61°43'$	DA				44
172	EG 218	$8^h54^m16^s$	$+40°27'9$	DA	$\log\theta = 0.28$		1.78	
	GD 98				$\log\theta' = 0.18$		2.08	38
173	EG 172							
	LP 90-70	8^h55^m8	$60°29'$	DCe?		$12^m.8$		3
	G 234-51				8700 K	$12^m.8$		5
174	EG 182							
	G 47-18	8^h56^m2	$33°09'$	4670		$12^m.7$		3
	G 51-32			4670		$13^m.50$		35
						$12^m.73$ ptm		
175	EG 219	$8^h58^m41^s$	$+36°19'0$	DA	$\log\theta = 0.53$		1.64; 2.02	
	GD 99				$\log\theta' = 0.42$		1.84+ 2.20	4
176	Gr 349							
	G 252-27	$9^h00^m25^s$	$73°27'$	DC?				5
	LP 36-115			DC		$14^m.92$ ptm		35
177	Gr 250			DCp	7.5×10^3 K		$R = 0.0101\, R_0$	50

tinued)

Name of star	1950	1950	Sp. type	T_{eff} or θ_e	M_v	$\log^R R_0$	References
(2)	(3)	(4)	(5)	(6)	(7)	(8)	(9)
G 195-19	$9^h12^m.5$	53°39′	DC?	7600 K			5
				8500 K			
			VDXp		$13^m.84$		35
					$13^m.57$		
					$13^m.62$		22
EG 64				9.4×10^3 K		$R = 0.0182\, R_0$	50
G 116-16	$9^h13^m.4$	44°12′	DAs		$11^m.7$		10
				0.55		1.75	9
EG 65				12.5×10^3 K		$R = 0.0142\, R_0$	50
G 117-B15A	$9^h21^m.2$	35°30′	DA		$13^m.4$		10
			DA		$11^m.33$		35
					$11^m.77$ ptm		
BPM 5639	$9^h28^m.6$	−71°20′	DA				44
Gr 324				8.3×10^3 K		$R = 0.0101\, R_0$	50
G 117-25	$9^h30^m46^s$	29°24′.8	DA		$13^m.28$		35
					$13^m.20$ ptm		
BPM 19652	$9^h34^m.3$	−58°43′	DA				44
EG 66							
LDS 275 AB	$9^h35^m.0$	−37°07′	DC		$13^m.0$		10
L 31							5
G 161-68	$9^h41^m18^s$	−6°49′.9	DAwk				2
EG 67				14.5×10^3 K		$R = 0.0170\, R_0$	50
G 116-52	$9^h43^m.5$	44°08′	DA		$11^m.2$		10
SA 29-130					$12^m.44$		
L 32					$10^m.44$		35
					$10^m.92$ ptm		
EG 68							
EG 117-B11B	$9^h43^m.5$	33°05′			$13^m.3$		10
Gr 251							
G 195-42	9^h46^m55	53°29′.2	DC?				5
BPM 19738	$9^h50^m.0$	−57°12′	DA				44
EG 69				8.7×10^3 K		$R = 0.0105\, R_0$	50
G 49-33	$9^h55^m.0$	24°47′	DA		$11^m.4$		10
				0.56	$13^m.2$	1.66; 2.02	6
				0.5		1.76; 2.12	
G 48-57	$9^h55^m39^s$	9°01′.3	DC?				2
			DAwk?				
BPM 6114	$9^h57^m.8$	−66°39′	DA				44
EG 252							
G 42-33	$9^h59^m08^s$	14°55′	DC?	7400 K			5
GD 111	$10^h00^m24^s$	43°02′.6	DA				1
EG 386							
GD 303	$10^h11^m17^s$	57°03′.5	DB				1
			DB		$10^m.99$ ptm		35
EG 183							
G 43-48	$10^h12^m.4$	8°12′.4	DF		$11^m.7$		3
					$14^m.0$		
BPM 19929	$10^h13^m.2$	−55°55′	DF				
EG 220	$10^h19^m33^s$	+46°16′.1	DAwk				38
GD 117							

(continued)

No.	Name of star	α_{1950}	δ_{1950}	Sp. type	T_{eff} or θ_e	M_v	log R/R_0	Refere[nces]
(1)	(2)	(3)	(4)	(5)	(6)	(7)	(8)	(9)
198	Gr 350							
	G 235-67	$10^h19^m38^s$	63°42′7	DAss		$13^m.73$		2
				DC?				2
				DAwk				
				DA				
199	G 43-54	$10^h26^m39^s$	11°42′9	DC	4600 K			2, 5
	HS							
200	GD 122	$10^h29^m16^s$	32°55′6	DC?	12 500 K			5
	WH							
201	EG 70							
	L 825	10^h31^m4	−11°29′	DAn		$10^m.6$		10
202	EG 71							
	F 34	10^h36^m7	43°22′	DO:		10^m0		10
					0.10		2.14	9
203	EG 72				7.2×10^3 K		$R = 0.0111\ R_0$	50
	G 44-32	10^h39^m2	14°32′			$11^m.4$		10
				DC	8000 K	$13^m.2$		5
				DA		$13^m.67$ ptm		35
204	BPM 6502	10^h42^m6	−69°02′	DA				44
205	EG 387							
	GD 124	$10^h46^m00^s$	−01°45′3	DB		$11^m.64$ ptm		35
206	EG 73							
	Ton 547	10^h46^m8	28°20′	DA		$11^m.3$		10
						$12^m.6$		
					0.52		1.76	
							2.02	
207	EG 221	$10^h52^m01^s$	+27°22′9	DAn	$\log\theta = 0.30$		1.70	
	Ton 556				$\log\theta' = 0.16$		2.12	4
	GD 125							
208	L 250-52	10^h53^m3	−55°04′	DAwk				44
209	G 163-28	$10^h55^m14^s$	−3°55′4	DF–DG				2
210	EG 74				8.1×10^3 K		$R = 0.0084\ R_0$	50
	L 898-25	10^h55^m1	−7°15′	DA		$11^m.4$		10
	L 34					$13^m.2$		
	LP 671-11			DA		13^m98		44
211	Gr 325							
	G 119-47	10^h56^m7	34°31′	DC	11 500 K	$11^m.97$ ptm		5
212	EG 75				19.2×10^3 K		$R = 0.0137\ R_0$	50
	LB 253	11^h04^m8	60°13′	DAn		10^m9		10
						$12^m.2$		
					0.39		1.74	9
							2.00	
213	EG 76				15.3×10^3 K		$R = 0.0126\ R_0$	50
	L 970-30	11^h05^m5	−4°53′	DA		11^m2		10
				DA		$11^m.23$		35
214	EG 77							
	GD 128	11^h07^m3	26°37′	DB		$10^m.0$		10
	Ton 573			DB		$11^m.65$ ptm		35
215	Gr 351							
	GD 105	$11^h08^m12^s$	56°21′0	DA		$12^m.92$ ptm		
	LP 129-10							
216	G 45-45	$11^h14^m06^s$	6°43′5	DAwk				2

tinued)

Name of star	α_{1950}	δ_{1950}	Sp. type	T_{eff} or θ_e	M_v	$\log R/R_0$	References
(2)	(3)	(4)	(5)	(6)	(7)	(8)	(9)
EG 78							
L 971-14	11^h15^m8	−2°58′	DC?		$11^m.0$		10
G 10-11					$12^m.6$		
				10 500 K			5
EG 79			DA–F	7.4×10^3 K		$R = 0.0123\,R_0$	50
R 627	11^h21^m7	21°39′	DF		$13^m.8$		10
L 36				0.61		2.10	9
			DF		$13^m.37$		35
			DAs		$13^m.57$ ptm		44
BPM 20912	11^h21^m5	−50°44′	DA				44
Gr 352							
GD 307	$11^h22^m30^s$	54°36′1	DA		$11^m.42$ ptm		35
LB 2012							
EG 80							
F 43	11^h26^m5	38°28′	DA		$10^m.6$		10
				0.28		1.81	9
					$10^m.49$ ptm		35
BPM 21022	11^h29^m3	−53°46′	DA				44
Gr 388							
G 147-65	$11^h33^m05^s$	35°50′9	DC				2
			M + DC				35
BPM 21065	11^h33^m5	−52°48′	DA				44
Gr 184							
GD 140	11^h34^m5	30°05′	DAwk		$10^m.8$		3
			DA		$10^m.45$ ptm		35
							38
EG 81							
F 46	11^h34^m9	14°27′	DO–B		$10^m.4$		10
EG 82							
L 145-141	11^h42^m9	−64°34′	4670		$13^m.0$		10
L 37							
Gr 353							
G 237-28	$11^h43^m03^s$	63°22′8	DFwk				2
LP 63-267			DC?				5
			DC–G		$14^m.65$ ptm		35
EG 83				16.5×10^3 K		$R = 0.0124\,R_0$	50
G 148-7	11^h43^m4	32°06′	DAs		$11^m.1$		10, 3
			DAn	$\log \theta = 0.47$		1.62; 1.90	
				$\log \theta' = 0.31$			
EG 84				10.5×10^3 K		$R = 0.0116\,R_0$	50
L 1405-40B	11^h47^m7	25°35′	DA		$12^m.4$		10
L 38							
Gr 354							
GD 312	$11^h49^m06^s$	41°03′8	DA		$11^m.46$		35
BPM 36430	11^h53^m7	−48°24′	DAwk				44
EG 85							
L 1261-24	11^h54^m2	18°39′	DC		$11^m.4$		10
G 57-29				8100 K	$13^m.1$		5
Gr 355							
G 255-2	$11^h59^m12^s$	80°21′7	DA		$12^m.21$ ptm		35
LP 7-200							
G 197-47	$12^h08^m58^s$	57°41′1	DA, Fwk				2

(continued)

No.	Name of star	α_{1950}	δ_{1950}	Sp. type	T_{eff} or θ_e	M_v	$\log R/R_0$	Refere
(1)	(2)	(3)	(4)	(5)	(6)	(7)	(8)	(9)
236	EG 86							
	HZ 21	12^h11^m4	33°12′	DO		$10^{m}.2$		10
				DO		$9^{m}.63$ ptm		35
237	EG 87							
	G 1	12^h13^m3	52°47′	(DA)		$13^{m}.7$		10
238	Gr 356							
	G 237-56	$12^h14^m42^s$	69°06′	DC				5
	LP 39-76					$14^{m}.05$ ptm		35
239	EG 88							
	L 1046-18B	12^h14^m3	3°14′	DAs		$12^{m}.5$		10
	G 13-26							
240	EG 89							
	G 148-B4B	12^h15^m0	32°22′			$13^{m}.3$		10
241	BPM 7543	12^h23^m8	−65°56′	DF				44
242	EG 90							
	HZ 28	12^h30^m0	41°16′	DA		$10^{m}.9$		10
						$12^{m}.0$		
243	EG 91							
	HZ 29	12^h32^m4	37°55′	DBp		$10^{m}.6$		10
				DB		$12^{m}.47$		35
244	G 186							
	G 148	12^h32^m5	47°54′	DA		$11^{m}.1$		3
						$13^{m}.0$		
				DA	$\log \theta = 0.47$		1.64; 2.02	
					$\log \theta' = 0.30$		1.94; 2.32	4
				DA				38
245	LFT 931	12^h36^m1	−49°33′	DA				44
246	EG 92				10.4×10^3 K		$R = 0.0099\, R_0$	50
	G 61-17	12^h44^m7	14°59′	DAs				10
					0.37		1.66	9
				DA		$12^{m}.61$		35
						$12^{m}.47$ ptm		
247	BPM 7855	12^h48^m4	−61°02′	DA				44
248	GD 151	$12^h49^m54^s$	18°13′0	DA				1
249	EG 93							
	HZ 34	12^h53^m0	37°49′	DOp		$10^{m}.3$		10, 5
250	Gr 357							
	GD 320	$12^h53^m11^s$	48°13′5	DA		$11^{m}.66$ ptm		35
251	BPM 21641	12^h53^m4	−56°09′	DF				44
252	EG 187							
	GD 153	12^h54^m6	22°18′	DAwk		$10^{m}.4$		3
	L 1336-41				$\log \theta = 0.17$		1.96	4
	LTT 13724				$\log \theta' = (0.10)$		2.20	38, 5
253	EG 94							
	L 1408-19	12^h57^m5	27°50′	DAs		$11^{m}.4$		10
	G 149-28					$13^{m}.3$		
254	BPM 7961	12^h57^m9	−72°18′	DA				44
255	EG 95				5.3×10^3 K		$R = 0.0096\, R_0$	50
	G 60-54	12^h57^m7	3°46′	DC		$15^{m}.4$		10
	W 457				5500 K	$15^{m}.4$		5
						$15^{m}.6$		
				DC		$14^{m}.66$ ptm		35

ntinued)

.	Name of star	α_{1950}	δ_{1950}	Sp. type	T_{eff} or θ_e	M_v	log R/R_0	References
	(2)	(3)	(4)	(5)	(6)	(7)	(8)	(9)
	EG 234	$12^h58^m30^s$	+59°20′2	DA				38
	GD 322							
	LB 2520							
	G 14-24	12^h59^m5	−1°49′	?		$12^m.2$		10
	EG 96							
	EG 97							
	HZ 39	13^h02^m4	28°24′	DAs		$10^m.4$		10
	EG 235	$13^h02^m30^s$	+59°43′5	DBp				38
	GD 323							
	LB 2539							
	Gr 389							
	G 61-29	$13^h03^m16^s$	18°17′0	DBe				10
				DBF		($12^m.45$)		35
	GD 268	$13^h04^m56^s$	15°00′5	DC?	15 000 K			5
	WH							
	G 256-7	$13^h09^m59^s$	85°18′5	DG				2
	LP 8-46							
	EG 98				60×10^3 K		$R = 0.0135\, R_0$	50
	HZ 43	13^h14^m0	29°22′	DAwk		$10^m.3$		10
				DAwk		$9^m.89$ ptm		35
						$9^m.00$		
	EG 188							
	G 177-31	13^h17^m1	45°21′	DAs	$\log \theta = 0.42$	$11^m.2$	1.82; 2.18	3
					$\log \theta' = 0.33$	$13^m.0$	1.92; 2.28	4
	L 258-46	13^h23^m0	−51°26′	DA				44
	Gr 358							
	G 199-71	$13^h25^m49^s$	58°10′4	DF		$13^m.94$ ptm		35
	LP 96-298							
	EG 99				12.8×10^3 K		$R = 0.0144\, R_0$	50
	G 14-58	13^h27^m7	−8°21′	DA		$11^m.0$		10
	W 485A					$10^m.8$		
						$11^m.33$		35
						$11^m.86$ ptm		
	Gr 391							
	G 165-7	$13^h28^m42^s$	30°45′3	DK				2
				DC		$13^m.33$ ptm		35
	Gr 359							
	GD 325	$13^h33^m59^s$	48°44′0	M + DB		12^m39 ptm		35
	G 165-18	$13^h34^m18^s$	36°39′1	DA, F?				2
	EG 100				4.1×10^3 K		$R = 0.0183\, R_0$	50
	G 62-53	13^h34^m3	3°58′	DK		$15^m.3$		10
	W 489					$15^m.02$		35
	L 49					$15^m.26$ ptm		
	EG 101				20.3×10^3 K		$R = 0.0083\, R_0$	50
	LDS 455A	13^h34^m3	−16°04′	(DA)?		$11^m.9$		10
	L 50							
	LFT 1025	13^h34^m6	−67°49′	DA				44
	EG 102							
	G 238-44	13^h37^m8	70°33′	DA		$10^m.8$		10
	Gw 70°5824					9.92		35
	L 52					$10^m.56$ ptm		

(continued)

No.	Name of star	α_{1950}	δ_{1950}	Sp. type	T_{eff} or θ_e	M_v	$\log R/R_0$	Referenc
(1)	(2)	(3)	(4)	(5)	(6)	(7)	(8)	(9)
275	EG 102							
	Ac 58°43662	13^h40^m6	57°15′	(DA)		$11^m.4$		10
						$13^m.2$		
				DA		$11^m.16$ ptm		35
276	Gr 360				7.6×10^3 K		$R = 0.0111\,R_0$	50
	G 63-54	$13^h44^m59^s$	10°36′6	DA, F				2
				DAwk		$13^m.60$		35
277	LP 380-5	$13^h45^m48^s$	23°49′6	DC		$15^m.26$		44
278	G 150-43	$13^h47^m28^s$	28°19′3	DAwk				2
279	EG 104							
	L 619-50	13^h48^m1	−27°18′	(DA)		$13^m.0$		10
	L 53							
280	EG 173				13.6×10^3 K		$R = 0.0100\,R_0$	50
	G 165-B5	13^h54^m9	34°03′	DA		$12^m.0$		3
281	Gr 392							
	GD 330	$14^h01^m55^s$	52°21′1	DC		$13^m.42$ ptm		35
282	Gr 272							
	G 64-43	$14^h03^m46^s$	−1°05′3	DB, DC				1
				DB		11^m90 ptm		35
283	BPM 38165	14^h07^m5	−47°31′	DA				44
284	EG 189				18.2×10^3 K		$R = 0.0123\,R_0$	50
	GD 163	14^h08^m3	32°23′	DA		$11^m.9$		3
	L 1484-18					$12^m.5$		
	LTT 14141				0.38		1.95	9
							2.07	38
285	EG 105							
	LP 439-354	14^h09^m5	15°47′			$+15^m.2$		10
286	EG 106							
	LP 439-356	14^h11^m5	15°44′	DA		$11^m.2$		10
	G 135-29					$12^m.5$		
287	Gr 326							
	G 166-14	$14^h13^m26^s$	23°10′6	DA				1
				DA		$10^m.95$ ptm		35
288	EG 107							
	F 93	14^h15^m3	13°16′	DAwk				10
					0.20		1.92	9
289	Gr 361							
	GD 335	$14^h19^m57^s$	35°09′	DC?				5
				DC		$12^m.18$ ptm		35
290	EG 108							
	Ton 197	14^h21^m5	31°50′	DAwk		$10^m.7$		10
					0.27		2.03	9
291	EG 222	$14^h22^m12^s$	+9°30′9	DA				38
	GD 165							
	L 14236							
292	EG 110							
	L 19-2	14^h25^m4	−81°07′	(DA)		$11^m.0$		10
	L 54							
293	Gr 362							
	G 223-63	$14^h26^m09^s$	61°24′	DF–DC				5
	LP 98-57			DF–C		13^m91 ptm		35
294	G 200-42	$14^h26^m38^s$	44°17′2	DF				2

ntinued)

Name of star	α_{1950}	δ_{1950}	Sp. type	T_{eff} or θ_e	M_v	log R/R_0	References
(2)	(3)	(4)	(5)	(6)	(7)	(8)	(9)
EG 236	$14^h29^m54^s$	+37°19′7	DA				38
GD 336							
Gr 363							
GD 337	14^h33^m6	53°48′5	K + DA		$10^m_.98$ ptm		35
EG 111							
G 66-32	14^h48^m4	7°47′	DA		$11^m_.1$		10
L 1126-68					$12^m_.5$		
			DA		$11^m_.85$ ptm		35
Gr 297							
GD 173	$14^h51^m17^s$	00°37′7	DA				1
			DA		$10^m_.45$ ptm		35
Gr 298	14^h55^m	+29°	DA, F	8.6×10^3 K		$R = 0.0118\,R_0$	50
EG 393							
G 256-18	$14^h59^m40^s$	82°08′6	DB		$11^m_.38$ ptm		35
EG 190							
GD 175	15^h03^m2	−7°03′	DAwk		$11^m_.6$		3
					$14^m_.0$		38
BPM 9323	15^h04^m9	−77°00′	DA				44
EG 364							
GD 340	$15^h08^m44^s$	63°43′8	DA		$12^m_.44$ ptm		35
EG 191							
GD 178	15^h09^m4	32°16′	DAn		$11^m_.2$		3
					$13^m_.0$		
				log $\theta = 0.31$		1.72; 1.96	4
						2.32	38
EG 112				9.3×10^3 K		$R = 0.0183\,R_0$	50
G 201-39	15^h10^m6	56°37′			$11^m_.5$		10
LP 135-155			DA		$11^m_.67$		35
					$12^m_.78$ ptm		
BPM 23251	15^h11^m6	−53°47′	DA–F				44
BPM 39481	15^h22^m0	−46°12′	DF				44
EG 237	$15^h25^m30^s$	+43°23′5	DA				38
GD 185	15^h31^m5	−2°17′	DAn		$11^m_.0$		3
BPM 77965					$12^m_.3$		
EG 192			DA		$11^m_.18$		35
							38
GD 347	$15^h34^m48^s$	50°23′7	DA		$13^m_.08$ ptm		35
EG 223	$15^h39^m37^s$	−3°32′1	DA				38
GD 189							
EG 193							
GD 190	15^h42^m1	18°16′	DBs		$10^m_.7$		3
			DB	log $\theta = 0.26$		1.84	4
				log $\theta' = 0.17$		(2.10)	
			DB				35
EG 113				39×10^3 K		$R = 0.0094\,R_0$	50
1°3129 B	15^h44^m2	0°53′	DAwk				10
				0.12		2.05	9
EG 114				8.1×10^3 K		$R = 0.0198\,R_0$	50
L 481-60	15^h44^m2	−37°45′	DAs		$12^m_.0$		10
−37°6571B							44
L 60							

(continued)

No.	Name of star	α_{1950}	δ_{1950}	Sp. type	T_{eff} or θ_e	M_v	log R/R_0	Referen
(1)	(2)	(3)	(4)	(5)	(6)	(7)	(8)	(9)
315	Gr 366							
	GD 352	$15^h50^m25^s$	62°36′	DC				5
				DC				35
316	EG 174				12.6×10^3 K		$R = 0.0154\ R_0$	50
	G 152-B4B	$15^h55^m.4$	−9°00′	DAs		$11^m.2$		3
317	EG 115				11.2×10^3 K		$R = 0.0111\ R_0$	50
	G 180-23	$15^h59^m.6$	36°58′	DA		$11^m.3$		10
	R 808					$12^m.8$		
	L 61			DA		$11^m.97$		35
318	EG 116							
	G 2	$16^h06^m.7$	42°13′	DA		$11^m.2$		10
						$12^m.6$		
					0.49		1.70	9
							1.98	
319	EG 117				9.6×10^3 K		$R = 0.0067\ R_0$	50
	G 138-8	$16^h09^m.1$	13°30′	DAs		$11^m.2$		10
						$12^m.7$		
						$13^m.82$		59
320	GD 196	$16^h10^m46^s$	16°39′.6	DA				1
321	Gr 194							
	GD 198	$16^h12^m.6$	−11°11′	DB		$10^m.8$		38
322	Gr 195							
	G 153-40	$16^h14^m.7$	−12°50′	DA		$11^m.1$		3
	LTT 6494					$12^m.8$		
323	EG 118							
	L 770-3	$16^h15^m.1$	−15°28′	DAwk		$10^m.5$		10
	G 153-41							
324	GD-38°10980	$16^h20^m.2$	−39°07′	DA		$10^m.6$		44
325	BPM 24107	$16^h20^m.6$	−53°37′	DA				44
326	G 202-49	$16^h24^m48^s$	47°45′.2	DA				2
327	Gr 327				6.9×10^3 K		$R = 0.0084\ R_0$	50
	G 138-31	$16^h25^m.5$	9°19′	DC?	7300 K			5
				DC				35
328	EG 119							
	G 180-57	$16^h26^m.8$	36°51′	DFp		$11^m.1$		10
	R 640					$12^m.8$		
	L 62							9
				DF		$12^m.65$		35
329	BPM 890	$16^h28^m.8$	−87°19′	DA				44
330	GD 354	$16^h30^m23^s$	61°48′.7					1
331	G 180-63	$16^h33^m25^s$	43°23′.8	DA-Fwk?				2
332	Gr 258				6.3×10^3 K		$R = 0.0117\ R_0$	50
	Gr 225-68	$16^h33^m28^s$	57°15′	DC	6500 K	$13^m.9$		5, 52
	LP 101-16			DC–F		$14^m.15$		35, 49
	CM Dra					$14^m.12$		
333	Gr 259							
	G 138-47	$16^h35^m22^s$	13°46′.7	DC	7300 K			5
334	GD 202	$16^h36^m25^s$	16°00′.1	DA				1
335	Gr 328							
	G 138-49	$16^h36^m28^s$	05°46′.7	DAs				35
336	EG 120				9.7×10^3 K		$R = 0.0131\ R_0$	50
	L 1491-27	$16^h37^m.4$	33°32′	DAs		$11^m.3$		10
						$13^m.0$		

tinued)

Name of star	α_{1950}	δ_{1950}	Sp. type	T_{eff} or θ_e	M_v	log R/R_0	References
(2)	(3)	(4)	(5)	(6)	(7)	(8)	(9)
EG 196							
G 138-56	$16^{h}39^{m}6$	15°19′	DAs		$11\overset{m}{.}5$		3
					$13\overset{m}{.}2$		
							35
Gr 329							
GD 356	$16^{h}39^{m}49^{s}$	53°46′9	DC				35
Gr 367							
G 257-38	$16^{h}41^{m}16^{s}$	73°16′0	DFwk				2
LP 43-146			DF–DC				5
			DC–F				35
			4670				
EG 238	$16^{h}41^{m}19^{s}$	38°46′7	DA				38
GD 357							
BPM 91679							
EG 121							
Ton 261	$16^{h}44^{m}5$	26°43′	DOp		$10\overset{m}{.}6$		10
EG 239	$16^{h}45^{m}25^{s}$	+32°33′7	DB				38
GD 358							
EG 368				11.8×10^3 K		$R = 0.0124\,R_0$	50
G 226-29	$16^{h}47^{m}38^{s}$	55°08′7	DA				35
EG 197				9.4×10^3 K		$R = 0.0136\,R_0$	50
G 169-34	$16^{h}55^{m}0$	21°32′	DAss	$\log \theta = 0.56$	$11\overset{m}{.}3$;	1.62; 1.96	3
				$\log \theta' = 0.43$	$13\overset{m}{.}0$	1.84; 2.18	4
			DA?	$\log \theta = 0.58$	$12\overset{m}{.}18$	1.65	
						1.99	9, 35
BPM 24601	$16^{h}59^{m}0$	−53°06′	DA		$11\overset{m}{.}3$	$R = 0.0157\,R_0$	50
G 139-13	$17^{h}05^{m}37^{s}$	3°01′6	DF–DG				2
				7.1×10^3 K		$R = 0.0104\,R_0$	50
EG 198				13.1×10^3 K		$R = 0.0102\,R_0$	50
G 181-B5B	$17^{h}07^{m}0$	33°17′	DA		$12\overset{m}{.}1$		3
				0.53		1.82	4
				0.43		2.00	4
			DA				35
EG 122							
L 845-70	$17^{h}08^{m}5$	−14°45′	DC?		$10\overset{m}{.}3$		10
L 66					$12\overset{m}{.}0$		
				10 200 K			5
EG 224	$17^{h}09^{m}50^{s}$	+23°04′7	DBn				38
GD 205							
BPM 92077							
BPM 24723	$17^{h}09^{m}8$	−57°34′	DAwk				44
EG 369							
G 240-47	$17^{h}10^{m}08^{s}$	68°23′4	DF		$14\overset{m}{.}94$		35
					$13\overset{m}{.}83$ ptm		
EG 370							
G 240-51	$17^{h}13^{m}28^{s}$	69°34′9	DA		$11\overset{m}{.}17$ ptm		35
GD 360	$17^{h}13^{m}44^{s}$	33°16′4	DA				1
BPM 24754	$17^{h}14^{m}8$	−54°44′	DA				44
EG 123				11.6×10^3 K		$R = 0.0148\,R_0$	50
W 672B	$17^{h}16^{m}2$	2°01′	DA		$12\overset{m}{.}2$		10
G 19-20							
BPM 24866	$17^{h}26^{m}1$	−57°53′	DB				

(continued)

No.	Name of star	α_{1950}	δ_{1950}	Sp. type	T_{eff} or θ_e	M_v	log R/R_0	Refere
(1)	(2)	(3)	(4)	(5)	(6)	(7)	(8)	(9)
357	G 227-5	$17^h28^m02^s$	56°00′5	DC				2
	HZ							
358	BPM 24960	17^h33^m2	−54°24′	DAs				44
359	EG 371				8.6×10^3 K		$R = 0.0132\, R_0$	50
	G 140-2	$17^h36^m14^s$	5°17′5	DA				2
				DA		$12^m.61$		35
						$13^m.04$		
360	EG 240	$17^h37^m01^s$	41°54′0	DA				38
	GD 363							
361	EG 175				10.6×10^3 K		$R = 0.0186\, R_0$	50
	G 154-B5	17^h43^m1	−13°17′	DA		$11^m.2$		3
362	BPM 25114	17^h43^m8	−52°06′	DAp				44
363	EG 372				5.3×10^3 K		$R = 0.0100\, R_0$	50
	G 240-72	$17^h48^m53^s$	70°52′7	DC				2
	LP 44-113			Dxp		$15^m.31$		35
						$14^m.44$ ptm		
						$14^m.49$		22
364	EG 124							
	G 140-B1B	17^h50^m5	9°49′	DC		$12^m.7$		4
					10500 K			5
365	EG 199				7.5×10^3 K		$R = 0.0135\, R_0$	50
	G 259-21	17^h57^m0	82°44′	DAss		$11^m.6$		3
	LP 9-231				0.71	$13^m.5$	1.54	9
						$13^m.18$		44
				DA		$13^m.03$		35
						$13^m.58$ ptm		
366	EG 241	$18^h09^m43^s$	28°29′0	DA				38
	GD 375							
	BPM 92960							
367	EG 373							
	G 206-17	$18^h11^m41^s$	32°47′7	DA–F		$13^m.45$ ptm		35
368	EG 330							
	G 206-18	18^h11^m8	37°47′9	DC?		$14^m.0$ ptm		5
								35
369	G 183-35	$18^h14^m06^s$	24°53′7	DC?				2
	HZ							
370	EG 176							
	G 155-15	18^h21^m3	−13°10′	DAwk		$11^m.5$		3
						$13^m.3$		38
371	Gr 288				6.4×10^3 K		$R = 0.0209\, R_0$	50
	G 141-2	$18^h18^m13^s$	12°37′4	DC				2
	HZ				5800 K			5
372	EG 242	$18^h22^m01^s$	41°02′2	DB				38
	GD 378							
373	EG 125							
	R 137	18^h24^m8	4°02′	DA		$11^m.2$		10
	G 21-15					$12^m.8$		
					0.48		1.74	9
							2.03	
374	EG 126				9.4×10^3 K		$R = 0.0139\, R_0$	50
	L 993-18	18^h26^m5	−4°31′	DAs		$11^m.2$		10

ntinued)

Name of star	α_{1950}	δ_{1950}	Sp. type	T_{eff} or θ_e	M_v	log R/R_0	References
(2)	(3)	(4)	(5)	(6)	(7)	(8)	(9)
G 21-16					$12^{m}.8$		
				0.58		1.63	9
						1.95	
EG 177							
G 155-19	$18^{h}27^{m}9$	$-10°39'$	DA		$11^{m}.2$		3
					$12^{m}.8$		
Gr 374							
G 227-35	$18^{h}29^{m}21^{s}$	$54°45'2$	DCpd				2
			DXp		$14^{m}.02$		35
					$14^{m}.07$		22
G 184-12	$18^{h}31^{m}49^{s}$	$19°43'7$	DC?				2
BPM 11593	$18^{h}34^{m}7$	$-78°09'$	DA				44
EG 200							
G 155-34	$18^{h}40^{m}2$	$-11°18'$	DAs		$11^{m}.4$		3
LTT 7421					$12^{m}.8$		
L 849-15?							
EG 225	$18^{h}40^{m}57^{s}$	$4°17'3$	DAss	$\log \theta_e = 0.56$		1.68; 1.98	4
GD 215				$\log \theta_e' = 0.43$		1.98	
EG 127							
L 1498-127	$18^{h}55^{m}7$	$33°52'$	DAn		$11^{m}.2$		10
				10.5×10^3 K	$12^{m}.8$	$R = 0.0126\, R_0$	50
EG 128							
G 141-54	$18^{h}57^{m}5$	$11°54'$	DAs		$11^{m}.3$		10
				0.55		1.67	9
						1.99	
			DA		$12^{m}.30$ ptm		35
G 205-52	$18^{h}58^{m}36^{s}$	$39°18'4$	DA				2
EG 129							
G 260-15	$19^{h}00^{m}6$	$70°36'$	4135		$12^{m}.3$		10
Gw 70°8247			DXp		$12^{m}.67$		35
L 73					$12^{m}.07$ ptm		
EG 130				11.4×10^3 K		$R = 0.0148\, R_0$	50
G 142-B2B	$19^{h}11^{m}3$	$13°31'$	DAn		$11^{m}.5$		10
				0.52		1.73	9
Gr 375				6.8×10^3 K		$R = 0.0095\, R_0$	50
G 125-3	$19^{h}17^{m}15^{s}$	$38°0'$	DC?				2
			DC?	6700 K	$14^{m}.3$		5
					$14^{m}.28$		35
					$13^{m}.89$ ptm		
EG 131				11.4×10^3 K		$R = 0.0107\, R_0$	50
GL 754.IA	$19^{h}17^{m}9$	$-7°45'$	DAwk		$12^{m}.6$		10
LDS 678B			DC?	11 000 K			5
L 74			DAwk		$12^{m}.19$		35
					$12^{m}.23$ ptm		
GD 218	$19^{h}18^{m}14^{s}$	$11°05'0$	DA				1
EG 331							
GD 533	$19^{h}18^{m}57^{s}$	$72°31'8$	DA		$10^{m}.14$ ptm		35
EG 201							
GD 219	$19^{h}19^{m}4$	$14°35'$	DAn		$11^{m}.1$		3
BPM 94172					$12^{m}.5$		
				0.48		1.67	9
						1.97	
			DA				38

(continued)

No.	Name of star	α_{1950}	δ_{1950}	Sp. type	T_{eff} or θ_e	M_v	log R/R_0	References
(1)	(2)	(3)	(4)	(5)	(6)	(7)	(8)	(9)
391	EG 132				17.2×10^3 K		$R = 0.0125\ R_0$	50
	LDS 683 B	19^h32^m9	−13°36′	DA		$11^m.1$		10
	L 75							
392	EG 277	19^h35^m	+27°	DA	10.5×10^3 K		$R = 0.0149\ R_0$	50
393	EG 226	$19^h36^m34^s$	32°46′5	DA				38
	BPM 94484							
	GD 222							
394	EG 133							
	L 1573-31	19^h40^m4	37°24′	DB		$10^m.7$		10
								9
				DB		$11^m.20$		35
395	EG 134				20.9×10^3 K		$R = 0.0113\ R_0$	50
	G 142-50	19^h43^m3	16°20′	DA		$10^m.9$		10
						$12^m.0$		
					0.34		1.80	9
							2.02	
396	GD 14462	19^h44^m2	−42°08′	Pec				44
397	Gr 394							
	GD 385	$19^h50^m22^s$	25°01′5	DA		$12^m.08$ ptm		35
398	BPM 12843	19^h53^m2	−71°31′	DA				44
399	EG 135				8.2×10^3 K		$R = 0.0111\ R_0$	50
	G 92-40	19^h54^m0	−1°09′	DAs		$13^m.1$		10
	L 997-21				0.56		1.98; 0.56	4
	L 79				0.45			
400	Gr 395							
	GD 387	$20^h03^m59^s$	43°45′7	DA		$11^m.64$		35
401	EG 136			WdB		$10^m.4$		57
	WZ Sge	20^h05^m3	17°33′	DAep		$11^m.0$		10
				DAe		$12^m.24$ ptm		35
402	EG 137							
	L 710-30	20^h07^m3	−21°55′	DA		$13^m.0$		10
	L 80							
403	Gr 376							
	G 230-30	$20^h08^m14^s$	51°03′3	DA		$11^m.07$ ptm		35
404	Gr 332							
	GD 543	$20^h09^m58^s$	62°16′7	DA		$10^m.97$		35
405	Gr 333							
	GD 229	$20^h10^m23^s$	31°04′6	Dxp		$11^m.16$		35
						$11^m.21$		22
406	EG 138				7.6×10^3 K		$R = 0.0099\ R_0$	50
	G 24-9	20^h11^m8	6°34′	DC?		$14^m.3$		10
				DC	7000 K	$13^m.9$		35
407	BPM 26734	20^h18^m3	−58°31′	DC?				44
408	Gr 396							
	GD 390	$20^h25^m52^s$	48°48′1	DA		$12^m.43^s$		35
409	EG 243							
	Gr 397							
	GD 391	$20^h28^m05^s$	39°03′4	DA		$10^m.08$		35
410	EG 139				21.5×10^3 K		$R = 0.0122\ R_0$	50
	G 186-31	20^h32^m3	24°33′	DA		$10^m.8$		10
	W 1346				0.35		1.76	9
	L 82			DA		$10^m.57$		35
						$10^m.64$		

ntinued)

Name of star	α_{1950}	δ_{1950}	Sp. type	T_{eff} or θ_e	M_v	log R/R_0	References
(2)	(3)	(4)	(5)	(6)	(7)	(8)	(9)
EG 140							
L 116-79	20^h30^m6	−68°16′	(DA)		$10^m.9$		10
GD 231	$20^h32^m58^s$	18°49′1		log $\theta = 0.36$		1.74; 2.12	4
				log $\theta_e = 0.25$		1.98; 2.36	
EG 141							
L 711-10	20^h39^m7	−20°16′	DA		$12^m.0$		10
BPM 13491	20^h39^m6	−68°16′	DA				44
BPM 13537	20^h43^m0	−63°31′	DA				44
G 187-8	$20^h48^m13^s$	26°19′6	DC				2
EG 202				3.4×10^3 K		$R = 0.0180\ R_0$	50
GL 812 B	20^h54^m1	−05°02′	DC		$15^m.7$		3
R 193 B				3700 K			5
			DC		$15^m.73$		35
Gr 398							
GD 392	$20^h58^m20^s$	34°14′5	DC		$12^m.15$		35
GD 393	$20^h58^m56^s$	50°39′4	DAs				1
Gr 262							
LTT 16151	20^h59^m7	31°37′	DC	10 500 K	$12^m.3$		5
G 187-15			DC		$12^m.31$		35
					$12^m.48$ ptm		
Gr 377				6.8×10^3 K		$R = 0.0140\ R_0$	50
G 144-51	$20^h59^m46^s$	19°01′2	DA, Fwk				2
			DAwk		$13^m.37$		35
					$13^m.83$ ptm		
G 187-16	$20^h59^m59^s$	24°45′7	DC				2
HS							
EG 142							
L 24-52	21^h05^m2	−82°01′	(DA)		$11^m.4$		
L 83					$13^m.0$		
Gr 334							
G 212-B1A	21^h08^m0	42°45′	DC				5
			DC		$11^m.77$		35
					$12^m.01$ ptm		
EG 244							
Gr 399							
GD 394	$21^h11^m03^s$	49°53′7	DO?		$10^m.17$		35
				9.3×10^3 K		$R = 0.0145\ R_0$	50
G 187-32	$21^h11^m34^s$	26°09′3	DAs				2
LTT 16224							
L 212-10	21^h15^m8	−56°03′	DA				44
Gr 400							
GD 547	$21^h16^m30^s$	67°32′1	DA		$11^m.21$		35
Gr 378							
G 231-40	$21^h17^m22^s$	53°59′9	DA		$10^m.97$		35
EG 143				14.1×10^3 K		$R = 0.0093\ R_0$	50
G 231-43	21^h24^m8	54°59′	(DA)		$11^m.1$		10
R 198					$12^m.5$		
L 85					$11^m.94$		35
					$11^m.48$		
EG 144				15.4×10^3 K		$R = 0.0153\ R_0$	50
G 261-43	21^h26^m6	73°25′	DA		$11^m.1$		10

(continued)

No.	Name of star	α_{1950}	δ_{1950}	Sp. type	T_{eff} or θ_e	M_v	log R/R_0	Referer
(1)	(2)	(3)	(4)	(5)	(6)	(7)	(8)	(9)
	73°8031				0.46	$12^{m}.5$	1.75	
							2.03	9
	L 86							
				DA		$10^{m}.94$		35
						$11^{m}.19$		
432	EG 145							
	G 26-10	$21^{h}29^{m}6$	−0°00′	DB		$12^{m}.0$		10
	LDS 749 B			DB		$11^{m}.83$		35
						$11^{m}.61$		
433	EG 146							
	GD 233	$21^{h}32^{m}0$	−4°46′	DB		$10^{m}.5$		10
	TT 8579							9
	L 1002-62							
434	EG 227	$21^{h}34^{m}18^{s}$	21°51′1	DA				38
	GD 234							
435	EG 147				17.6×10^3 K		$R = 0.0158\, R_0$	50
	82°3818	$21^{h}36^{m}7$	82°49′	DA		$11^{m}.0$		10
	L 88					$12^{m}.5$		
436	G 126-18	$21^{h}36^{m}28^{s}$	22°55′7	DA				1
437	G 126-25	$21^{h}39^{m}43^{s}$	13°15′0	DA				2
438	EG 148				8.6×10^3 K		$R = 0.0140\, R_0$	50
	G 126-27	$21^{h}40^{m}3$	20°45′	DC		$10^{m}.9$		10
	L 1363-3					$12^{m}.2$		
				DC?	9500 K	$12^{m}.6$		5
				DC		$12^{m}.57$		35
						$12^{m}.98$		
439	GD 396	$21^{h}43^{m}16^{s}$	35°18′3	DA				1
440	EG 149							
	G 26-31	$21^{h}44^{m}9$	−7°58′	DB		$10^{m}.7$		10
	L 930-80							9
				DB		$11^{m}.31$		35
						$11^{m}.21$		39
441	G 188-27	$21^{h}47^{m}39^{s}$	28°03′0	DC				2
	H							
442	G 232-38	$21^{h}48^{m}14^{s}$	53°54′5	DA, F				2
443	Gr 401							
	GD 397	$21^{h}49^{m}56^{s}$	37°12′1	DA		$11^{m}.70$		35
444	EG 150				18.3×10^3 K		$R = 0.0131\, R_0$	50
	G 93-48	$21^{h}49^{m}9$	2°02′	DA		$11^{m}.0$		10
						$12^{m}.0$		
445	Gr 402							
	GD 398	$21^{h}50^{m}19^{s}$	33°53′6	DA		$10^{m}.73$		35
446	EG 151							
	G 93-53	$21^{h}51^{m}5$	−1°31′	DA		$11^{m}.4$		10
	L 1003-16					$13^{m}.5$		
					0.62		1.61	9
							2.03	
447	BPM 27606	$21^{h}53^{m}5$	−51°14′	DC?		$13^{m}.9$		44
448	EG 152							
	LDS 766A	$21^{h}54^{m}9$	−43°42′	(DA)		$10^{m}.5$		10
	L 90							
449	BPM 44347	$22^{h}03^{m}9$	−48°34′	DA				44

inued)

Name of star	α_{1950}	δ_{1950}	Sp. type	T_{eff} or θ_e	M_v	log R/R_0	References
(2)	(3)	(4)	(5)	(6)	(7)	(8)	(9)
Gr 302				8.4×10^3 K		$R = 0.0090\, R_0$	50
G 18-34	$22^{h}07^{m}21^{s}$	14°14′9	DA				1
					$13\overset{m}{.}31$		35
L 119-34	$22^{h}16^{m}2$	−65°44′	DF				44
EG 153							
LDS 785 A	$22^{h}24^{m}6$	−34°27′	DB		$10\overset{m}{.}5$		10
							9
EG 228	$22^{h}26^{m}38^{s}$	6°07′1	DA				38
GD 236							
BPM 27891	$22^{h}32^{m}0$	−57°31′	DA				58
Gr 379							
G 233-19	$22^{h}34^{m}36^{s}$	52°48′	DC		$11\overset{m}{.}92$		5
			DC				35
EG 229	$22^{h}40^{m}09^{s}$	−4°30′0	DAwk				38
GD 240							
PHL 380							
Feige 106							
EG 154							
G 28-13	$22^{h}40^{m}5$	−1°43′	DA, F		$11\overset{m}{.}4$		10
					$13\overset{m}{.}5$		
				0.60		1.63	9
						1.97	
			DA–F		$12\overset{m}{.}87$		35
EG 155				10.5×10^3 K		$R = 0.0085\, R_0$	50
G 67-23	$22^{h}46^{m}6$	22°20′	DA		$11\overset{m}{.}0$		10
					$12\overset{m}{.}7$		
			DA		$12\overset{m}{.}83$		35
					$12\overset{m}{.}37$		
Gr 283				5.6×10^3 K		$R = 0.0160\, R_0$	50
G 128-7	$22^{h}48^{m}57^{s}$	29°23′7	DAs—F?		$13\overset{m}{.}99$		35
					$14\overset{m}{.}48$		
LP 701-29	$22^{h}51^{m}09^{s}$	−7°02′3	DK pec		$16^{m}13$		44
EG 178				7.4×10^3 K		$R = 0.0139\, R_0$	50
G 156-64	$22^{h}53^{m}2$	−8°06′	DA, F		$13\overset{m}{.}2$		3
			DA–F		$13\overset{m}{.}85$		35
EG 230	$22^{h}53^{m}12^{s}$	−6°16′9	DB				38
GD 243			DBn				47
EG 156							
G 28-27	$22^{h}54^{m}9$	7°40′	4670		$12\overset{m}{.}4$		3
EG 231	$22^{h}54^{m}16^{s}$	+12°36′8	DA				38
GD 244							
EG 232	$22^{h}56^{m}22^{s}$	+24°59′7	DA				38
GD 245							
Gr 403							
GD 404	$23^{h}02^{m}25^{s}$	45°43′2	DA		$10\overset{m}{.}56$		35
EG 233	$23^{h}09^{m}50^{s}$	+10°30′7	DAwk				38
GD 246							
BPM 97859							
Gr 404							
GD 556	$23^{h}11^{m}50^{s}$	55°12′1	DA		$11\overset{m}{.}92$		35
				6.6×10^3 K		$R = 0.0112\, R_0$	50
G 157-35	$23^{h}12^{m}43^{s}$	−02°26′3	DF				2

(continued)

No.	Name of star	α_{1950}	δ_{1950}	Sp. type	T_{eff} or θ_e	M_v	log R/R_0	Referen
(1)	(2)	(3)	(4)	(5)	(6)	(7)	(8)	(9)
470	EG 157							
	F 108	$23^{h}13^{m}6$	−2°07′	DA		$10\overset{m}{.}5$		10
					0.14		2.09	9
471	Gr 264							
	PHL 459	$23^{h}17^{m}1$	−17°22′	DC				5
472	EG 158							
	F 110	$23^{h}17^{m}4$	−5°26′	DOp		$10\overset{m}{.}4$		10
473	Gr 303							
	GD 559	$23^{h}19^{m}15^{s}$	69°09′9	DA		$10\overset{m}{.}54$		35
474	Gr 380							
	G 128-62	$23^{h}23^{m}28^{s}$	25°36′0	DA, F				2
				DA		$14\overset{m}{.}41$		
475	Gr 335							
	GD 248	$23^{h}23^{m}36^{s}$	15°43′8	DC	10 600 K			5
				DC		$12\overset{m}{.}40$		35
476	EG 159				11.9×10^3 K		$R = 0.0094\ R_0$	
	G 29-38	$23^{h}26\overset{m}{.}6$	4°58′	DA		$11\overset{m}{.}2$		10
				? DA		$12\overset{m}{.}5$		
						$12\overset{m}{.}24$		35
						$11\overset{m}{.}65$		
477	EG 160							
	G 3	$23^{h}29^{m}2$	40°45′	DAn		$11\overset{m}{.}0$		10
						$12\overset{m}{.}6$		
					0.44		1.72	9
							2.04	
478	EG 161				9.7×10^3 K		$R = 0.0124\ R_0$	50
	G 128-72	$23^{h}29^{m}5$	26°42′	DAn		$11\overset{m}{.}2$		10
						$12\overset{m}{.}8$		
479	GD 251	$23^{h}31^{m}53^{s}$	29°02′1	DA				1
480	Gr 336							
	PC 233712	$23^{h}37^{m}42^{s}$	12°21′3	DA		$(12\overset{m}{.}16)$		35
481	EG 162				12.5×10^3 K		$R = 0.0134\ R_0$	50
	G 130-5	$23^{h}41^{m}4$	32°15′	DA		$12\overset{m}{.}4$		10
	1512-34B				0.52		1.91	9
	L 93			DA		$11\overset{m}{.}64$		35
						$11\overset{m}{.}70$		
482	Gr 304							
	GD 561	$23^{h}42^{m}57^{s}$	80°40′2	DA		$9\overset{m}{.}66$		35
483	Gr 405							
	G 30-20	$23^{h}47^{m}20^{s}$	12°49′5	DA		$12\overset{m}{.}05$		35
484	G 130-15	$23^{h}47^{m}23^{s}$	29°17′7	DA, F				2
485	EG 163							
	R 826A	$23^{h}51^{m}5$	−33°33′	(DA)		$12\overset{m}{.}5$		10
	L94							
486	EG 164							
	505-42	$23^{h}51^{m}7$	−36°50′	(DA)		$12\overset{m}{.}5$		10
	L 95							
487	EG 165				18.7×10^3 K		$R = 0.0040\ R_0$	50
	L 96	$23^{h}59^{m}6$	−43°25′	(DAs)		$13\overset{m}{.}4$		10
488	G 171-27	$23^{h}52^{m}24^{s}$	40°11′1	DC				5

White Dwarfs. Photometry

M_{pg}	V	$B-V$	$U-B$	I	R	G	B	U	References
(10)	(11)	(12)	(13)	(14)	(15)	(16)	(17)	(18)	(19)
$14^m.5$	$14^m.90$	$0^m.465$	$-0^m.435$						5
$15^m.1$									
	$14^m.30$			$14^m.92$	$14^m.68$	$14^m.14$	$14^m.08$	$14^m.02$	35
$14^m.5$		$-0^m.29$	$-1^m.21$						38, 46
	$16^m.85$	$+0^m.25$	$-0^m.70$						10
	$16^m.77$			$16^m.87$	$16^m.84$	$16^m.81$	$16^m.87$	$17^m.08$	35
$14^m.3$									2
	$14^m.46$			$14^m.30$	$14^m.35$	$14^m.59$	$14^m.79$	$15^m.20$	35
	$15^m.31$	$+0^m.24$	$-0^m.58$						10
	$15^m.22$	$-0^m.12$	$-0^m.97$						10
	$15^m.27$			$15^m.90$	$15^m.68$	$15^m.07$	$14^m.94$	$14^m.97$	35
	$16^m.47$			$16^m.66$	$16^m.67$	$16^m.46$	$16^m.48$	$16^m.85$	35
$16^m.$									36
$15^m.7$	$14^m.87$			$14^m.43$	$14^m.55$	$15^m.13$	$15^m.52$	$16^m.26$	35
	$15^m.52$	$+0^m.24$	$-0^m.68$						10
	$15^m.56$			$15^m.96$	$15^m.85$	$15^m.50$	$15^m.57$	$15^m.85$	35
$14^m.0$		$-0^m.22$	$-1^m.20$						
	$16^m.14$			$16^m.38$	$16^m.43$	$16^m.04$	$16^m.01$	$16^m.05$	35
$13^m.4$	$14^m.1$								5
$14^m.1$									
$15^m.1$	$14^m.60$	$0^m.65$	$-0^m.32$						5
	$14^m.53$	$0^m.62$	$-0^m.14$						
$14^m.5$	$(14^m.5)$								5
$15^m.00$									
	$14^m.48$			$15^m.19$	$15^m.06$	$14^m.46$	$14^m.37$	$14^m.36$	35
	$12^m.36$	$+0^m.56$	$+0^m.04$						10
	$12^m.41$			$12^m.30$	$12^m.31$	$12^m.50$	$12^m.80$	$13^m.45$	35
	$14^m.20$	$-0^m.015$	$-0^m.79$						44
	$13^m.36$	$-0^m.25$	$-1^m.07$						
$15^m.5$									2
	$15^m.26$	$+0^m.35$	$-0^m.50$						10
	$14^m.10$	$+0^m.14$	$-0^m.50$						10
$12^m.27$	$14^m.0$			$14^m.12$	$14^m.10$	$14^m.03$	$14^m.16$	$14^m.45$	35
									36
	$13^m.57$	$-0^m.29$	$-1^m.11$						
	$15^m.26$	$+0^m.20$	$-0^m.58$						
	$15^m.25$	$-0^m.23$	$-1^m.05$						10
	$14^m.93$	$+0^m.09$	$-0^m.64$						10
	$13^m.82$	$+0^m.11$	$-0^m.80$						10
			$-0^m.77$						
$13^m.6$									5
$13^m.9$									5
$14^m.4$									
$14^m.9$	$14^m.94$	$+0^m.40$		$14^m.97$	$14^m.92$	$15^m.03$	$15^m.18$	$15^m.67$	35, 5
$14^m.0$									46
	$14^m.10$	$+0^m.20$	$-0^m.54$						10

(continued)

No.	M_{pg}	V	$B-V$	$U-B$	I	R	G	B	U	References
	(10)	(11)	(12)	(13)	(14)	(15)	(16)	(17)	(18)	(19)
35		$13^m.06$			$13^m.81$	$13^m.55$	$12^m.83$	$12^m.68$	$12^m.94$	35
36		$12^m.84$	$+0^m.34$	$-0^m.50$						10
		$12^m.83$				$12^m.78$	$12^m.91$	$13^m.05$	$13^m.40$	35
37										36
38		$17^m.41$	$-0^m.05$	$-0^m.62$						10
	$16^m.2$									5
	$16^m.9$									
	$17^m.3$									
39		$15^m.05$	$+0^m.25$	$-0^m.65$						10
40	$12^m.2$	$12^m.35$	$-0^m.05$							5
		$12^m.33$			$12^m.61$	$12^m.50$	$12^m.17$	$12^m.10$	$11^m.89$	35
41		$14^m.40$			$15^m.17$	$14^m.89$	$14^m.12$	$13^m.94$	$13^m.80$	35
42	$13^m.6$	$13^m.0$								47
43		$17^m.16$	$+0^m.21$	$-0^m.81$						10
44		$14^m.88$			$15^m.40$	$15^m.26$	$14^m.73$	$14^m.75$	$15^m.37$	35
45	$14^m.5$									
46		$13^m.22$	$-0^m.04$	$-0^m.85$						10
47		$14^m.47$	$0^m.24$		$14^m.46$	$14^m.46$	$14^m.57$	$14^m.71$	$15^m.02$	35
48		$14^m.80$			$15^m.02$	$14^m.94$	$14^m.79$	$14^m.87$	$15^m.17$	35
49		$16^m.23$	$0^m.73$	$+0^m.02$						10
		$16^m.3$	$0^m.56$							
	$15^m.90$									5
		$16^m.30$			$15^m.85$	$15^m.97$	$16^m.53$	$16^m.91$	$17^m.84$	35
50		$13^m.68$	$-0^m.92$	$+0^m.16$						10
51		$15^m.83$	$+0^m.07$	$-0^m.71$						10
52		$12^m.65$	$-0^m.06$	$-0^m.83$						10
53										
54	$14^m.0$		$0^m.21$	$-0^m.68$						38
55		$12^m.25$	$-0^m.23$	$-1^m.25$						10
		$12^m.39$			$12^m.34$	$12^m.52$	$12^m.19$	$12^m.07$	$11^m.75$	35
56		$13^m.73$			$14^m.37$	$14^m.18$	$13^m.55$	$13^m.49$	$13^m.75$	35
57	$16^m.1$									2
58	$16^m.0$									2
59		$14^m.08$	$+0^m.23$	$-0^m.59$						
60	$16^m.0$									2
61			$-0^m.03$	$-1^m.06$						1
62			$-0^m.05$	$-0^m.92$						1
		$15^m.48$			$16^m.05$	$15^m.84$	$15^m.38$	$15^m.30$	$15^m.30$	35
63		$14^m.07$	$-0^m.11$	$-1^m.06$						
64		$11^m.2$								10
65		$15^m.54$	$1^m.32$	$0^m.38$						10
		$15^m.38$			$13^m.36$	$14^m.39$	$16^m.18$	$17^m.10$	$18^m.09$	35
66	$13^m.0$		$0^m.08$	$-0^m.68$						38
67	$16^m.1$									2
68	$16^m.3$									2
	$16^m.6$									5
		$16^m.85$			$16^m.92$	$16^m.85$	$16^m.92$	$16^m.99$	$17^m.38$	35
69		$14^m.0$								10
70		$15^m.50$	$0^m.21$	$-0^m.55$						10

tinued)

M_{pg}	V	$B-V$	$U-B$	I	R	G	B	U	References
(10)	(11)	(12)	(13)	(14)	(15)	(16)	(17)	(18)	(19)
	$15^{m}.20$	$0^{m}.30$	$-0^{m}.52$						10
	$15^{m}.30$			$15^{m}.20$	$15^{m}.20$	$15^{m}.45$	$15^{m}.53$	$15^{m}.84$	35
	$15^{m}.00$	$+0^{m}.21$	$-0^{m}.66$						44
	$15^{m}.74$	$-0^{m}.06$	$-0^{m}.77$						60
	$16^{m}.52$	$-0^{m}.20$	$-1^{m}.10$						10
	$16^{m}.57$			$17^{m}.30$	$17^{m}.06$	$16^{m}.35$	$16^{m}.23$	$16^{m}.03$	35
	$14^{m}.47$	$0^{m}.15$	$-0^{m}.67$						10
	$14^{m}.54$				$14^{m}.93$	$14^{m}.35$	$14^{m}.36$	$14^{m}.70$	35
									39
	$15^{m}.52$	$1^{m}.83$	$1^{m}.14$						10
	$15^{m}.47$			$13^{m}.97$	$14^{m}.65$	$16^{m}.28$	$18^{m}.35$		35
	$13^{m}.80$	$0^{m}.10$	$-0^{m}.58$						10
					$14^{m}.25$	$13^{m}.77$	$13^{m}.77$	$14^{m}.16$	35
	$15^{m}.35$	$0^{m}.09$	$-0^{m}.65$						10
	$15^{m}.26$				$15^{m}.63$	$15^{m}.10$	$15^{m}.08$	$15^{m}.37$	35
	$14^{m}.14$	$0^{m}.17$	$-0^{m}.58$						10
	$13^{m}.86$	$-0^{m}.05$	$-0^{m}.88$						10
	$14^{m}.03$				$14^{m}.48$	$13^{m}.79$	$13^{m}.71$	$13^{m}.78$	35
	$15^{m}.65$	$1^{m}.05$	$0^{m}.32$						10
	$9^{m}.52$	$0^{m}.03$	$-0^{m}.68$						10
	$9^{m}.50$			$10^{m}.05$	$9^{m}.87$	$9^{m}.30$	$9^{m}.29$	$9^{m}.52$	35
	$15^{m}.63$	$0^{m}.16$	$-0^{m}.53$						10
	$16^{m}.62$	$0^{m}.14$	$-0^{m}.57$						10
	$15^{m}.32$	$-0^{m}.105$	$-1^{m}.08$						44
	$14^{m}.29$	$-0^{m}.02$	$-0^{m}.84$						10
$15^{m}.4$	$15^{m}.7$								5
$15^{m}.7$									
	$14^{m}.02$	$-0^{m}.09$	$-0^{m}.97$						10
$13^{m}.0$	$12^{m}.45$	$0^{m}.33$	$-0^{m}.49$						10
	$12^{m}.37$			$12^{m}.29$	$12^{m}.31$	$12^{m}.45$	$12^{m}.60$	$12^{m}.80$	35
	$13^{m}.95$	$0^{m}.33$	$-0^{m}.69$						10
	$13^{m}.88$			$12^{m}.58$	$13^{m}.55$	$13^{m}.97$	$14^{m}.04$	$14^{m}.33$	35
	$14^{m}.18$	$-0^{m}.03$	$-0^{m}.89$						10
	$14^{m}.23$				$14^{m}.73$	$14^{m}.02$	$13^{m}.91$	$14^{m}.01$	35
	$15^{m}.94$	$0^{m}.65$	$-0^{m}.06$						10
$15^{m}.8$									5
	$15^{m}.78$	$0^{m}.65$							59
		$-0^{m}.09$	$-0^{m}.99$						1
				$15^{m}.52$	$15^{m}.28$	$14^{m}.68$	$14^{m}.62$	$14^{m}.55$	35
	$14^{m}.10$	$0^{m}.14$	$-0^{m}.65$						10
	$14^{m}.83$			$15^{m}.40$	$15^{m}.23$	$14^{m}.71$	$14^{m}.63$	$14^{m}.53$	35
	$13^{m}.82$	$-0^{m}.15$	$-1^{m}.04$						10
	$13^{m}.86$				$14^{m}.40$	$13^{m}.60$	$13^{m}.48$	$13^{m}.35$	35
$16^{m}.0$									2
	$15^{m}.95$			$16^{m}.10$	$16^{m}.05$	$15^{m}.96$	$16^{m}.04$	$16^{m}.31$	35
$16^{m}.6$									2

(continued)

No.	M_{pg}	*V*	*B* − *V*	*U* − *B*	*I*	*R*	*G*	*B*	*U*	Referenc
	(10)	(11)	(12)	(13)	(14)	(15)	(16)	(17)	(18)	(19)
103	$15^m.0$		$0^m.23$	$-0^m.61$						38
104		$15^m.98$			$16^m.45$	$16^m.27$	$15^m.83$	$15^m.86$	$16^m.28$	35
105		$11^m.75$			$12^m.54$	$12^m.30$	$11^m.46$	$11^m.27$	$10^m.95$	35
106										
107		$16^m.20$	$0^m.31$	$-0^m.64$						10
108		$14^m.64$			$14^m.67$	$14^m.67$	$14^m.70$	$14^m.83$	$17^m.10$	35
109		$16^m.83$			$17^m.27$	$17^m.12$	$16^m.69$	$16^m.67$	$17^m.10$	35 36
110		$15^m.63$			$15^m.95$	$15^m.81$	$15^m.60$	$15^m.75$	$16^m.04$	35 36
111		$14^m.48$			$14^m.35$	$14^m.37$	$14^m.67$	$15^m.08$	$15^m.23$	35
112	$13^m.0$		$-0^m.25$	$-1^m.16$						38
113		$15^m.86$	$-0^m.05$	$-0^m.76$						10
		$15^m.92$	$-0^m.12$	$-0^m.74$						
	$15^m.5$ $16^m.0$									5
114		$14^m.52$	$1^m.06$	$0^m.81$						10
		$14^m.65$			$14^m.07$	$14^m.22$	$15^m.11$	$15^m.68$	$17^m.40$	35 36
115	$14^m.5$ $14^m.7$ $15^m.4$	$14^m.12$	$0^m.62$	$-0^m.12$						5
		$14^m.11$			$13^m.74$	$13^m.88$	$14^m.38$	$14^m.69$	$15^m.41$	35
116		$14^m.39$			$13^m.92$	$14^m.07$	$14^m.65$	$14^m.96$	$15^m.62$	35
117	$16^m.4$									2
118										
119	$15^m.0$		$0^m.00$	$-0^m.82$						38
120		$13^m.40$	$-0^m.15$	$-0^m.98$						10
121		$14^m.09$	$-0^m.09$	$-0^m.95$						44
122	$15^m.9$									2
123		$15^m.38$	$0^m.26$	$-0^m.73$						44
124		$16^m.58$	$0^m.21$	$-0^m.72$						10
125										
126	$16^m.1$									2
127	$15^m.0$		$0^m.13$	$-0^m.64$						38
128		$16^m.70$	$-0^m.05$	$-0^m.95$						10
129		$8^m.3$	$(0^m.12)$	(-1.03)						10
130		$12^m.10$	$-0^m.08$	$-0^m.92$						10
		$12^m.00$			$12^m.71$	$12^m.49$	$11^m.75$	$11^m.66$	$11^m.72$	35
131	$16^m.3$									2
132	$14^m.5$		$-0^m.21$	-1.19						5
133		$16^m.62$			$16^m.35$	$16^m.42$	$16^m.87$	$17^m.12$	$17^m.65$	35 36
134	$16^m.0$	$16^m.16$	$+0^m.59$	$-0^m.04$						5
135		$16^m.15$	$0^m.13$	$-0^m.69$						3, 10
	$15^m.9$									
136		$14^m.46$	$0^m.22$	$-0^m.72$						44
137		$15^m.64$	$0^m.30$	$-0^m.55$						10
138	$15^m.0$		$0^m.19$	$-0^m.52$						38

ntinued)

M_{pg}	V	$B-V$	$U-B$	I	R	G	B	U	References
(10)	(11)	(12)	(13)	(14)	(15)	(16)	(17)	(18)	(19)
$15^m.5$	$15^m.19$	$0^m.08$	$-0^m.79$						5
$14^m.5$		$-0^m.11$	$-0^m.99$						38
	$14^m.62$	$0^m.99$	$0^m.40$						10
$15^m.6$									5
$15^m.7$	$14^m.74$			$14^m.26$	$14^m.14$	$15^m.16$	$15^m.65$	$16^m.85$	35
$16^m.4$	$16^m.33$	$0^m.91$							5
	$16^m.33$			$15^m.85$	$15^m.97$	$16^m.67$	$17^m.09$	$18^m.09$	35
									10
	$14^m.16$	$+0^m.11$	$-0^m.66$						44
	$10^m.7$								10
	$12^m.98$	$0^m.29$	$-0^m.61$						10
	$12^m.91$			$12^m.98$	$12^m.93$	$12^m.94$	$13^m.08$	$13^m.27$	35
	$15^m.06$	$-0^m.16$	$-0^m.79$						44
	$15^m.62$			$16^m.22$	$15^m.89$	$15^m.55$	$15^m.70$	$15^m.85$	35
	$16^m.3$	$0^m.09$							5
	$15^m.66$			$15^m.59$	$15^m.56$	$15^m.65$	$15^m.80$	$16^m.15$	35
	$13^m.66$	$-0^m.13$	$-0^m.95$						10
$14^m.9$	$15^m.06$	$0^m.12$							5
$15^m.2$									
$15^m.4$									35
	$15^m.06$			$15^m.35$	$15^m.24$	$15^m.01$	$15^m.03$	$15^m.05$	
$16^m.0$									5
	$16^m.08$			$16^m.3$	$16^m.09$	$16^m.11$	$16^m.24$	$16^m.61$	35
	$14^m.5$	$0^m.4$							10
$14^m.6$	$14^m.09$	$0^m.66$	$-0^m.17$						5
	$14^m.5$								
	$13^m.56$	$-0^m.02$	$-0^m.79$						10
	$15^m.56$	$-0^m.05$	$-0^m.89$						44
	$15^m.58$			$16^m.00$	$15^m.83$	$15^m.50$	$15^m.50$	$15^m.54$	35
	$13^m.02$	$+0^m.05$	$-0^m.90$						44
	$16^m.01$			$16^m.66$	$16^m.48$	$15^m.85$	$15^m.96$	$16^m.16$	35
	$15^m.60$			$16^m.28$	$16^m.00$	$15^m.71$	$15^m.91$	$16^m.10$	35
	$16^m.58$	$0^m.36$	$-0^m.58$						10
$16^m.0$		$0^m.20$	$-0^m.52$						38
	$17^m.85$	$0^m.16$	$-0^m.68$						10
	$18^m.23$	$0^m.05$	$-0^m.47$						10
	$17^m.66$	$0^m.04$	$-0^m.60$						10
	$12^m.0$								10
	$15^m.55$	$-0^m.06$	$-0^m.93$						10
	$15^m.61$			$16^m.10$	$16^m.01$	$15^m.49$	$15^m.38$	$15^m.37$	35
	$15^m.70$			$15^m.70$	$15^m.67$	$15^m.73$	$15^m.89$	$16^m.19$	35
	$14^m.73$	$-0^m.02$	$-0^m.89$						44
$15^m.0$		$-0^m.13$	$-0^m.94$						38
	$16^m.38$	$0^m.21$	$-0^m.64$						3
$15^m.5$									5
$15^m.9$									

(continued)

No.	M_{pg}	V	$B-V$	$U-B$	I	R	G	B	U	Referenc
	(10)	(11)	(12)	(13)	(14)	(15)	(16)	(17)	(18)	(19)
174		$15^m.18$	$0^m.00$	$-0^m.89$						3
		$15^m.14$			$15^m.39$	$15^m.25$	$15^m.22$	$15^m.03$	$15^m.07$	35
175	$15^m.0$		$0^m.19$	$-0^m.59$						38
176	$16^m.8$	$17^m.0$	$(+0^m.76)$							5
	$17^m.2$	16:98			$16^m.46$	$16^m.56$	$17^m.32$	$17^m.76$	$18^m.79$	36
177	$14^m.8$	$13^m.79$	$0^m.30$	$-0^m.66$						5
		$13^m.93$	$0^m.35$	$-0^m.48$						
		$13^m.84$			$13^m.94$	$13^m.84$	$13^m.94$	$14^m.12$	$14^m.41$	35
178		$15^m.32$	$0^m.24$	$-0^m.53$						10
179		$15^m.52$	$0^m.20$	$-0^m.56$						10
		$15^m.35$			$15^m.90$	$15^m.70$	$15^m.25$	$15^m.25$	$15^m.65$	35
180		$15^m.44$	$0^m.175$	$-0^m.44$						44
181		$15^m.89$			$16^m.03$	$15^m.97$	$15^m.94$	$16^m.02$	$16^m.28$	35
182		$15^m.46$	$0^m.00$	$-0^m.92$						44
183	$14^m.3$	$15^m.0$								5
	$15^m.1$									
	$14^m.9$									
184	$16^m.4$									2
185		$13^m.32$	$0^m.07$	$-0^m.54$						10
		$13^m.20$			$13^m.81$	$13^m.63$	$13^m.00$	$12^m.93$	$13^m.45$	35
186		$17^m.24$	$0^m.40$	$-0^m.51$						10
187	$15^m.3$	$15^m.3$	0							5
188		$14^m.94$	$+0^m.20$	$-0^m.58$						44
189		$15^m.09$	$0^m.25$	$-0^m.54$						10
190	$17^m.3$									2
191		$14^m.60$	$-0^m.20$	$-0^m.98$						44
192	$15^m.1$	$15^m.37$	$0^m.36$	$-0^m.49$						5
193			$-0^m.08$	$-0^m.87$						1
194		$14^m.57$			$15^m.11$	$14^m.91$	$14^m.35$	$14^m.26$	$14^m.18$	35
195		$16^m.16$	$0^m.41$	$-0^m.40$						3
196		$15^m.10$	$+0^m.68$	$-0^m.12$						44
197	$16^m.5$		$0^m.33$	$-0^m.63$						38
198	$14^m.9$	$14^m.7$	$0^m.17$							2
		$14^m.66$			$14^m.59$	$14^m.59$	$14^m.76$	$14^m.92$	$15^m.29$	35
199	$17^m.1$									2
		$16^m.48$		$0^m.26$						5
200	$15^m.5$	$16^m.07$	$0^m.00$	$-0^m.88$						1
										5
201		$12^m.97$	$-0^m.15$	$-1^m.02$						10
202		$11^m.12$	$-0^m.30$	$-1^m.35$						10
203		$16^m.55$	$0^m.29$	$-0^m.58$						10
	$15^m.4$									5
		$16^m.86$			$16^m.73$	$16^m.75$	$16^m.90$	$17^m.08$	$17^m.35$	35
204		$13^m.09$	$-0^m.04$	$-0^m.89$						44
205			$-0^m.03$	$-0^m.88$						1
		$15^m.78$			$16^m.26$	$16^m.00$	$15^m.64$	$15^m.55$	$15^m.54$	35
206		$15^m.40$	$0^m.16$	$-0^m.58$						10
207	$14^m.0$		$-0^m.08$	$-0^m.98$						38
208		$14^m.32$	$-0^m.095$	$-0^m.62$						44

(continued)

M_{pg}	V	$B-V$	$U-B$	I	R	G	B	U	References
(10)	(11)	(12)	(13)	(14)	(15)	(16)	(17)	(18)	(19)
$16^m.0$									2
	$14^m.28$	$0^m.32$	$-0^m.51$						10
$15^m.4$	$15^m.58$	$-0^m.11$	$-0^m.70$						5
$15^m.5$	$15^m.93$	$-0^m.08$							
	$15^m.93$			$16^m.36$	$16^m.19$	$15^m.84$	$15^m.79$	$15^m.80$	35
	$13^m.80$	$-0^m.02$	$-0^m.81$						10
	$12^m.92$	$0^m.09$	$-0^m.69$						10
	$12^m.91$			$13^m.55$	$13^m.34$	$12^m.75$	$12^m.74$	$13^m.06$	35
	$15^m.0$								10
	$15^m.89$			$16^m.23$	$16^m.20$	$15^m.78$	$15^m.62$	$15^m.71$	35
	$16^m.90$			$17^m.40$	$17^m.05$	$16^m.90$	$16^m.99$	$17^m.38$	35
$16^m.9$									2
	$15^m.30$	$0^m.09$	$-0^m.76$						10
$14^m.9$									5
$15^m.0$									
$15^m.5$									
	$14^m.24$	$0^m.31$	$-0^m.52$						10
	$14^m.12$			$14^m.17$	$14^m.10$	$14^m.22$	$14^m.35$	$14^m.63$	35
	$14^m.86$	$0^m.07$	$-0^m.74$						58
	$15^m.35$			$15^m.96$	$15^m.71$	$15^m.19$	$15^m.20$	$15^m.54$	35
	$14^m.89$	$-0^m.13$	$-1^m.0$						10
	$14^m.92$			$15^m.62$	$15^m.38$	$14^m.67$	$14^m.61$	$14^m.52$	35
	$15^m.69$	$-0^m.12$	$-0^m.93$						44
$16^m.1$									2
	$15^m.63$			$13^m.59$	$14^m.82$	$16^m.37$	$17^m.01$	$17^m.59$	35
	$15^m.8$	$0^m.12$	$-0^m.65$						44
	$12^m.50$	$-0^m.06$	$-0^m.98$						3
	$12^m.38$			$13^m.18$	$12^m.87$	$12^m.17$	$12^m.06$	$12^m.06$	35
	$13^m.24$	$-0^m.30$	$-1^m.13$						10
	$11^m.44$	$0^m.19$	$-0^m.59$						10
	$11^m.50$	$0^m.18$	$-0^m.63$						44
$16^m.3$		$+0^m.60$							2
$15^m.0$	$16^m.4$								5
	$16^m.38$			$15^m.94$	$16^m.04$	$16^m.64$	$16^m.99$	$17^m.75$	35
	$13^m.68$	$0^m.05$	$-0^m.66$						10
	$15^m.55$	$0^m.23$	$-0^m.58$						10
	$16^m.06$			$16^m.60$	$16^m.41$	$15^m.86$	$15^m.93$	$16^m.22$	35
	$12^m.85$	$-0^m.205$	$-1^m.01$						44
	$15^m.54$	$0^m.30$	$-0^m.60$						10
$15^m.4$									5
$15^m.7$									
$16^m.0$									
	$16^m.04$			$16^m.40$	$16^m.30$	$15^m.90$	$16^m.05$	$16^m.44$	35
$15^m.9$									2
	$14^m.22$	$-0^m.36$	$-1^m.25$						10
	$14^m.58$			$15^m.36$	$15^m.12$	$14^m.32$	$14^m.08$	$13^m.18$	35
	$13^m.34$	$0^m.53$	$-0^m.48$						10
$16^m.6$	$17^m.0$	$(0^m.32)$							5
	$17^m.08$			$17^m.00$	$16^m.94$	$17^m.26$	$17^m.48$	$17^m.93$	35

(continued)

No.	M_{pg}.	V	$B-V$	$U-B$	I	R	G	B	U	References
	(10)	(11)	(12)	(13)	(14)	(15)	(16)	(17)	(18)	(19)
239		$14^{m}.9$								10
240		$16^{m}.95$	$0^{m}.38$	$-0^{m}.46$						10
241		$13^{m}.97$	$+0^{m}.40$	$-0^{m}.54$						44
242		$15^{m}.72$	$-0^{m}.04$	$-0^{m}.82$						10
243		$14^{m}.18$	$-0^{m}.23$	$-1^{m}.01$						10
		$14^{m}.11$			$14^{m}.77$	$14^{m}.47$	$13^{m}.92$	$13^{m}.75$	$13^{m}.69$	35
244		$14^{m}.52$	$0^{m}.06$	$-0^{m}.68$						3
245		$13^{m}.96$	$+0^{m}.18$	$-0^{m}.70$						44
246		$15^{m}.86$	$0^{m}.22$	$-0^{m}.53$						10
		$15^{m}.83$			$16^{m}.15$	$16^{m}.05$	$15^{m}.75$	$15^{m}.86$	$16^{m}.21$	35
247		$15^{m}.9$	$-0^{m}.14$	$-1^{m}.07$						44
248			$-0^{m}.08$	$-0^{m}.85$						1
249		$15^{m}.66$	$-0^{m}.28$	$-1^{m}.26$						10
250		$16^{m}.46$			$17^{m}.03$	$16^{m}.84$	$16^{m}.35$	$16^{m}.35$	$16^{m}.71$	35
251		$15^{m}.01$	$0^{m}.68$	$-0^{m}.165$						44
252		$13^{m}.42$	$-0^{m}.25$	$-1^{m}.18$						3
253		$15^{m}.40$	$0^{m}.28$	$-0^{m}.58$						10
254		$15^{m}.18$	$+0^{m}.01$	$-0^{m}.73$						44
255		$15^{m}.90$	$0^{m}.64$	$-0^{m}.09$						10
	$16^{m}.0$									5
	$16^{m}.4$									
	$16^{m}.8$									35
		$15^{m}.89$			$15^{m}.44$	$15^{m}.57$	$16^{m}.20$	$16^{m}.46$	$17^{m}.25$	
256	$15^{m}.0$									47
257		$12^{m}.78$	$0^{m}.75$	$0^{m}.0$						10
258		$15^{m}.44$	$-0^{m}.32$	$-1^{m}.16$						10
259	$14^{m}.5$									47
260			$-0^{m}.10$	$-0^{m}.97$						3
		$15^{m}.98$			$16^{m}.14$	$16^{m}.08$	$15^{m}.89$	$15^{m}.85$	$15^{m}.65$	35
261	$16^{m}.0$	$17^{m}.06$	$-0^{m}.05$	$-0^{m}.99$						
262	$16^{m}.3$									2
263		$12^{m}.86$	$-0^{m}.10$	$-1^{m}.14$						10
		$12^{m}.98$			$13^{m}.68$	$13^{m}.50$	$12^{m}.70$	$12^{m}.53$	$12^{m}.22$	35
264		$14^{m}.13$	$0^{m}.03$	$-0^{m}.56$						3
265		$14^{m}.60$	$0^{m}.00$	$-0^{m}.79$						44
266		$16^{m}.71$			$16^{m}.60$	$16^{m}.55$	$16^{m}.85$	$17^{m}.02$	$17^{m}.40$	35
267,		$12^{m}.30$	$0^{m}.08$	$-0^{m}.61$						10
		$12^{m}.20$			$12^{m}.71$	$12^{m}.51$	$12^{m}.11$	$12^{m}.08$	$12^{m}.40$	35
268	$16^{m}.0$									2
		$16^{m}.04$			$16^{m}.02$	$15^{m}.95$	$15^{m}.98$	$17^{m}.25$	$18^{m}.59$	35
269		$13^{m}.97$			$13^{m}.50$	$13^{m}.96$	$13^{m}.94$	$13^{m}.93$	$13^{m}.94$	35
270	$15^{m}.9$									2
271		$14^{m}.68$	$0^{m}.96$	$0^{m}.37$						10
		$14^{m}.68$			$13^{m}.96$	$14^{m}.14$	$15^{m}.12$	$15^{m}.60$	$16^{m}.84$	35
272		$15^{m}.35$	$-0^{m}.06$	$-0^{m}.83$						10
273		$15^{m}.67$	$+0^{m}.30$	$-0^{m}.12$						44
274		$12^{m}.79$	$-0^{m}.09$	$-0^{m}.84$						10
		$12^{m}.76$			$13^{m}.45$	$13^{m}.22$	$12^{m}.52$	$12^{m}.46$	$12^{m}.53$	35

tinued)

M_{pg}	V	$B-V$	$U-B$	I	R	G	B	U	References
(10)	(11)	(12)	(13)	(14)	(15)	(16)	(17)	(18)	(19)
	$16^{m}.73$	$0^{m}.55$	$-0^{m}.27$						10
	$13^{m}.34$			$13^{m}.96$	$13^{m}.74$	$13^{m}.77$	$13^{m}.12$	$13^{m}.55$	35
$15^{m}.7$									2
	$15^{m}.06$			$15^{m}.09$	$15^{m}.09$	$15^{m}.17$	$15^{m}.33$	$15^{m}.66$	35
	$15^{m}.65$	$1^{m}.10$							
$16^{m}.3$									2
	$15^{m}.0$	$0^{m}.1$							10
	$16^{m}.16$	$0^{m}.08$	$-0^{m}.70$						3
	$16^{m}.40$			$16^{m}.39$	$16^{m}.39$	$16^{m}.45$	$16^{m}.55$	$16^{m}.79$	35
		$-0^{m}.05$	$-0^{m}.99$						1
	$15^{m}.90$			$16^{m}.26$	$16^{m}.09$	$15^{m}.71$	$15^{m}.76$	$15^{m}.69$	35
	$14^{m}.31$	$-0^{m}.07$	$-0^{m}.89$						44
	$14^{m}.06$	$-0^{m}.03$	$-0^{m}.81$						3
	$18^{m}.12$	$0^{m}.96$	$0^{m}.30$						10
	$16^{m}.20$	$0^{m}.07$	$-0^{m}.65$						10
		$-0^{m}.10$	$-0^{m}.95$						1
	$16^{m}.41$			$17^{m}.05$	$16^{m}.87$	$16^{m}.24$	$16^{m}.15$	$16^{m}.17$	35
	$15^{m}.33$	$-0^{m}.23$	$-1^{m}.12$						10
$16^{m}.0$	$17^{m}.0$	$(-0^{m}.29)$							35
	$16^{m}.95$			$17^{m}.30$	$17^{m}.10$	$16^{m}.81$	$16^{m}.86$	$16^{m}.74$	35
	$15^{m}.30$	$-0^{m}.12$	$-1^{m}.00$						10
$14^{m}.5$		$0^{m}.14$	$-0^{m}.59$						38, 47
	$13^{m}.0$	$0^{m}.0$							10
$16^{m}.5$	$17^{m}.0$	$(0^{m}.25)$							5
$16^{m}.8$									5
	$17^{m}.00$			$16^{m}.91$	$16^{m}.91$	$17^{m}.17$	$17^{m}.30$	$17^{m}.65$	
$16^{m}.5$									2
$14^{m}.5$									47
	$16^{m}.11$			$16^{m}.08$	$16^{m}.40$	$15^{m}.86$	$15^{m}.85$	$15^{m}.86$	35
	$15^{m}.47$	$0^{m}.02$	$-0^{m}.67$						10
	$15^{m}.58$			$16^{m}.02$	$15^{m}.86$	$15^{m}.38$	$15^{m}.54$	$15^{m}.88$	35
		$-0^{m}.14$	$-0^{m}.96$						1
	$15^{m}.20$			$15^{m}.85$	$15^{m}.61$	$14^{m}.95$	$14^{m}.82$	$14^{m}.81$	35
	$14^{m}.78$			$15^{m}.31$	$15^{m}.15$	$14^{m}.64$	$14^{m}.55$	$14^{m}.49$	35
	$15^{m}.91$	$0^{m}.40$	$-0^{m}.42$						3
	$15^{m}.17$	$0^{m}.04$	$-0^{m}.91$						44
	$14^{m}.71$			$15^{m}.09$	$14^{m}.97$	$14^{m}.63$	$14^{m}.79$	$15^{m}.10$	35
	$14^{m}.11$	$0^{m}.09$	$-0^{m}.65$						3
	$16^{m}.24$	$0^{m}.23$	$-0^{m}.60$						10
	$16^{m}.26$			$16^{m}.55$	$16^{m}.44$	$16^{m}.33$	$16^{m}.26$	$16^{m}.63$	35
	$14^{m}.04$	$+0^{m}.77$	$0^{m}.00$						44
$15^{m}.0$									47
	$13^{m}.68$	$+0^{m}.68$	$-0^{m}.10$						44
	$14^{m}.03$	$0^{m}.00$	$-0^{m}.81$						3
	$13^{m}.90$			$14^{m}.51$	$14^{m}.33$	$13^{m}.73$	$13^{m}.70$	$13^{m}.83$	35
	$15^{m}.81$			$15^{m}.90$	$15^{m}.87$	$15^{m}.80$	$15^{m}.89$	$16^{m}.19$	35
$15^{m}.5$		$0^{m}.25$	$-0^{m}.55$						38,47

(continued)

No.	M_{pg}	V	$B-V$	$U-B$	I	R	G	B	U	Reference
	(10)	(11)	(12)	(13)	(14)	(15)	(16)	(17)	(18)	(19)
312		$14^{m}.72$	$-0^{m}.10$	$-1^{m}.00$						3
		$14^{m}.63$			$15^{m}.36$	$15^{m}.05$	$14^{m}.43$	$14^{m}.29$	$14^{m}.26$	35
313		$15^{m}.27$	$-0^{m}.32$	$-1^{m}.19$						10
314		$12^{m}.80$	$0^{m}.30$	$-0^{m}.40$						10
315	$16^{m}.5$	$17^{m}.3$								5
		$16^{m}.78$			$16^{m}.92$	$16^{m}.85$	$16^{m}.78$	$16^{m}.89$	$17^{m}.00$	35
316		$14^{m}.80$	$0^{m}.09$	$-0^{m}.63$						3
317		$14^{m}.36$	$0^{m}.17$	$-0^{m}.56$						10
		$14^{m}.33$			$14^{m}.77$	$14^{m}.62$	$14^{m}.24$	$14^{m}.25$	$14^{m}.71$	35
318		$13^{m}.85$	$0^{m}.06$	$-0^{m}.54$						10
319		$15^{m}.10$	$0^{m}.23$	$-0^{m}.65$						10
320										
321		$15^{m}.53$	$-0^{m}.11$	$-0^{m}.94$						30
322		$16^{m}.14$	$0^{m}.03$	$-0^{m}.68$						3
323		$12^{m}.40$	$-0^{m}.25$	$-1^{m}.02$						10
324		$11^{m}.0$	$+0^{m}.145$	$-0^{m}.955$						44
325		$15^{m}.64$	$-0^{m}.03$	$-0^{m}.69$						44
326	$16^{m}.0$									2
327	$15^{m}.7$	$16^{m}.14$	$0^{m}.39$	$-0^{m}.50$						5
		$16^{m}.16$								
		$16^{m}.16$			$16^{m}.08$	$16^{m}.08$	$16^{m}.29$	$16^{m}.48$	$16^{m}.89$	35
328		$13^{m}.86$	$0^{m}.18$	$-0^{m}.68$						10
		$13^{m}.79$			$14^{m}.05$	$13^{m}.94$	$13^{m}.79$	$13^{m}.83$	$13^{m}.97$	35
329		$14^{m}.58$	$0^{m}.22$	$-0^{m}.63$						44
330										
331	$15^{m}.4$									2
332		$14^{m}.00$	$0^{m}.49$	$-0^{m}.36$						5
		$15^{m}.7$	$0^{m}.31$							
		$15^{m}.00$			$14^{m}.76$	$14^{m}.80$	$15^{m}.15$	$15^{m}.35$	$15^{m}.86$	35
333	$16^{m}.5$	$16^{m}.90$	$0^{m}.41$	$-0^{m}.50$						5
334										
335		$16^{m}.46$			$16^{m}.54$	$16^{m}.55$	$16^{m}.48$	$16^{m}.57$	$16^{m}.87$	35
336		$14^{m}.64$	$0^{m}.22$	$-0^{m}.58$						10
337		$15^{m}.66$	$0^{m}.35$	$-0^{m}.57$						3
		$15^{m}.66$			$15^{m}.65$	$15^{m}.66$	$15^{m}.70$	$15^{m}.78$	$16^{m}.10$	35
338		$15^{m}.05$			$15^{m}.13$	$15^{m}.07$	$15^{m}.10$	$15^{m}.23$	$15^{m}.51$	35
339	$16^{m}.3$									2
		$16^{m}.5$								5
		$16^{m}.50$			$16^{m}.74$	$16^{m}.68$	$16^{m}.51$	$16^{m}.52$	$16^{m}.71$	35
340	$15^{m}.0$									47
341		$16^{m}.5$	$-0^{m}.26$	$-0^{m}.97$						10
342	$14^{m}.0$									47
343		$12^{m}.17$			$12^{m}.70$	$12^{m}.48$	$12^{m}.05$	$12^{m}.04$	$12^{m}.43$	35
344		$14^{m}.08$	$0^{m}.24$	$-0^{m}.62$						3
		$14^{m}.06$			$14^{m}.31$	$14^{m}.18$	$14^{m}.00$	$14^{m}.10$	$14^{m}.44$	35
345		$13^{m}.47$	$+0^{m}.10$	$-0^{m}.70$						44
346	$15^{m}.2$									2
347		$15^{m}.92$	$0^{m}.17$	$-0^{m}.55$						3
		$15^{m}.83$			$16^{m}.31$	$16^{m}.17$	$15^{m}.67$	$15^{m}.75$	$16^{m}.15$	35

(continued)

M_{pg}	V	$B-V$	$U-B$	I	R	G	B	U	References
(10)	(11)	(12)	(13)	(14)	(15)	(16)	(17)	(18)	(19)
	$14^m.30$	$0^m.02$	$-0^m.68$						10
$14^m.3$									5
$15^m.0$		$-0^m.09$	$-0^m.98$						38
	$15^m.10$	$+0^m.03$	$-0^m.46$						44
	$17^m.49$			$17^m.35$	$17^m.35$	$17^m.59$	$17^m.75$	$18^m.21$	35
	$13^m.27$			$13^m.85$	$13^m.65$	$13^m.04$	$13^m.09$	$13^m.39$	35
	$15^m.55$	$0^m.27$	$-0^m.59$						44
	$14^m.26$	$0^m.13$	$-0^m.59$						10
	$15^m.27$	$-0^m.04$	$-1^m.02$						44
$16^m.21$									2
	$15^m.8$	$0^m.46$	$-0^m.45$						44
$15^m.5$									2
	$15^m.80$			$15^m.89$	$15^m.97$	$15^m.82$	$15^m.88$	$16^m.20$	35
$15^m.0$									47
	$14^m.22$	$0^m.32$	$-0^m.45$						3
	$15^m.74$	$0^m.09$	$-0^m.99$						44
$14^m.5$									2
	$14^m.21$			$13^m.54$	$13^m.69$	$14^m.34$	$14^m.49$	$15^m.07$	35
	$14^m.15$	$0^m.41$							44
	$15^m.72$	$0^m.10$	$-0^m.78$						4
$16^m.0$									5
	$14^m.51$	$0^m.41$	$-0^m.48$						3
	$14^m.25$			$14^m.25$	$14^m.25$	$14^m.35$	$14^m.47$	$14^m.81$	35
$15^m.0$									47
	$16^m.39$			$16^m.40$	$16^m.35$	$17^m.18$	$17^m.41$	$17^m.82$	35
$17^m.0$	$17^m.05$	$0^m.49$							5
	$17^m.1$	$(0^m.27)$							
	$17^m.07$			$16^m.83$	$16^m.91$	$17^m.18$	$17^m.41$	$17^m.82$	35
$16^m.4$									2
$15^m.7$									2
	$15^m.89$		$-0^m.17$						5
$14^m.0$									38
	$15^m.60$	$0^m.42$	$-0^m.60$						3
	$13^m.90$	$0^m.05$	$-0^m.55$						4
	$14^m.54$	$0^m.24$	$-0^m.56$						10
	$14^m.25$	$0^m.14$	$-0^m.63$						3
$16^m.1$									2
	$15^m.53$			$15^m.45$	$15^m.42$	$15^m.75$	$15^m.87$	$16^m.33$	35
$16^m.3$									2
	$15^m.45$	$-0^m.06$	$-0^m.80$						44
	$14^m.18$	$0^m.15$	$-0^m.61$						3
$15^m.0$		$0^m.25$	$-0^m.62$						38
	$14^m.64$	$0^m.17$	$-0^m.60$						10
	$15^m.52$	$0^m.20$	$-0^m.57$						10
	$15^m.55$			$15^m.86$	$15^m.75$	$15^m.44$	$15^m.50$	$15^m.92$	35
$15^m.7$									2
	$13^m.9$	$0^m.05$	$-0^m.85$						10
									5
	$13^m.10$			$13^m.52$	$13^m.45$	$13^m.04$	$13^m.04$	$13^m.01$	35

(continued)

No.	M_{pg}	V	$B-V$	$U-B$	I	R	G	B	U	Reference
	(10)	(11)	(12)	(13)	(14)	(15)	(16)	(17)	(18)	(19)
384		$14^{m}.00$	$0^{m}.12$	$-0^{m}.60$						10
385	$15^{m}.0$									2
	$15^{m}.0$	$14^{m}.54$								5
		$14^{m}.58$								
		$14^{m}.6$								
		$14^{m}.64$			$14^{m}.50$	$14^{m}.53$	$14^{m}.75$	$14^{m}.96$	$15^{m}.40$	
386		$12^{m}.24$	$0^{m}.07$	$-0^{m}.84$						10
	$12^{m}.2$									5
387		$12^{m}.24$			$12^{m}.61$	$12^{m}.46$	$12^{m}.17$	$12^{m}.16$	$12^{m}.16$	35
388			$-0^{m}.11$	$-0^{m}.86$						1
389		$15^{m}.12$			$15^{m}.81$	$15^{m}.52$	$14^{m}.82$	$14^{m}.70$	$14^{m}.82$	35
390		$13^{m}.01$	$0^{m}.06$	$-0^{m}.66$						3
391	$13^{m}.5$		$-0^{m}.12$	$-0^{m}.88$						38
392		$15^{m}.95$	$0^{m}.05$	$-0^{m}.76$						10
393		$14^{m}.51$	$-0^{m}.09$	$-0^{m}.97$						10
		$14^{m}.52$			$15^{m}.15$	$14^{m}.92$	$14^{m}.41$	$14^{m}.25$	$14^{m}.21$	35
394		$14^{m}.08$	$-0^{m}.06$	$-0^{m}.86$						10
395		$10^{m}.4$	$0^{m}.0$	$-0^{m}.8$						44
396		$15^{m}.13$			$15^{m}.54$	$15^{m}.39$	$14^{m}.96$	$15^{m}.14$	$15^{m}.49$	35
397		$15^{m}.15$	$0^{m}.01$	$-0^{m}.99$						44
398		$13^{m}.69$	$0^{m}.25$	$-0^{m}.60$						10
399		$16^{m}.41$			$16^{m}.83$	$16^{m}.73$	$16^{m}.23$	$16^{m}.25$	$16^{m}.49$	35
400		$15^{m}.2$	$0^{m}.10$	$-0^{m}.72$						10
		$15^{m}.32$			$15^{m}.62$	$15^{m}.52$	$15^{m}.19$	$15^{m}.26$	$15^{m}.20$	35
401		$14^{m}.2$								10
402		$16^{m}.21$			$16^{m}.84$	$16^{m}.65$	$16^{m}.05$	$15^{m}.97$	$16^{m}.46$	35
403		$15^{m}.15$			$15^{m}.47$	$15^{m}.60$	$14^{m}.91$	$14^{m}.82$	$14^{m}.79$	35
404		$14^{m}.78$			$15^{m}.48$	$15^{m}.19$	$14^{m}.45$	$14^{m}.89$	$14^{m}.20$	35
405		$15^{m}.67$	$0^{m}.40$	$-0^{m}.44$						10
	$16^{m}.0$									5
406		$15^{m}.55$	$0^{m}.29$	$-0^{m}.56$						44
407		$15^{m}.85$			$16^{m}.20$	$16^{m}.06$	$15^{m}.75$	$15^{m}.91$	$16^{m}.25$	35
408		$13^{m}.84$			$14^{m}.41$	$13^{m}.84$	$13^{m}.06$	$12^{m}.98$	$12^{m}.96$	35
409		$11^{m}.54$	$-0^{m}.07$	$-0^{m}.87$						10
		$11^{m}.47$			$12^{m}.17$	$11^{m}.90$	$11^{m}.25$	$11^{m}.16$	$11^{m}.22$	35
410		$13^{m}.25$	$-0^{m}.06$	$-0^{m}.81$						10
411	$15^{m}.5$									47
412		$12^{m}.0$								10
413		$13^{m}.53$	$0^{m}.11$	$-0^{m}.78$						44
414		$15^{m}.52$	$-0^{m}.16$	$-1^{m}.09$						44
415	$16^{m}.3$									2
416		$16^{m}.68$	$1^{m}.13$	$0^{m}.64$						3
	$17^{m}.8$	$16^{m}.68$								5
	$18^{m}.1$									
		$16^{m}.70$			$15^{m}.80$	$16^{m}.10$	$17^{m}.30$	$17^{m}.94$	$19^{m}.30$	35
417		$15^{m}.71$			$16^{m}.05$	$15^{m}.88$	$15^{m}.62$	$15^{m}.56$	$15^{m}.64$	35
418										
419	$14^{m}.9$	$15^{m}.04$	$0^{m}.07$	$-0^{m}.75$						5
	$15^{m}.2$	$15^{m}.07$	$0^{m}.10$	$-0^{m}.79$						

(continued)

M_{pg}	V	$B-V$	$U-B$	I	R	G	B	U	References
(10)	(11)	(12)	(13)	(14)	(15)	(16)	(17)	(18)	(19)
$15^m.9$	$15^m.5$								
	$15^m.04$			$15^m.35$	$15^m.22$	$15^m.00$	$14^m.99$	$15^m.10$	35
$15^m.8$									2
	$16^m.37$			$16^m.22$	$16^m.26$	$16^m.48$	$16^m.63$	$17^m.03$	35
$16^m.3$									2
	$13^m.50$	$0^m.27$	$-0^m.59$						10
$15^m.7$	$15^m.9$	$0^m.05$							5
	$15^m.90$			$16^m.31$	$16^m.21$	$15^m.81$	$15^m.80$	$15^m.79$	35
	$13^m.06$			$13^m.75$	$12^m.57$	$12^m.81$	$12^m.66$	$12^m.38$	35
$15^m.5$									2
	$14^m.28$	$+0^m.26$	$-0^m.59$						44
	$16^m.14$			$16^m.80$	$16^m.55$	$15^m.97$	$15^m.97$	$16^m.49$	35
	$12^m.26$			$12^m.89$	$12^m.69$	$12^m.07$	$12^m.01$	$12^m.42$	35
	$14^m.66$	$0^m.15$	$-0^m.68$						10
	$14^m.63$			$15^m.15$	$15^m.00$	$14^m.48$	$14^m.45$	$14^m.85$	35
	$12^m.88$	$0^m.01$	$-0^m.66$						10
	$12^m.68$			$13^m.29$	$13^m.12$	$12^m.52$	$12^m.48$	$12^m.82$	35
	$14^m.73$	$-0^m.07$	$-0^m.92$						10
	$14^m.68$			$15^m.16$	$15^m.03$	$14^m.56$	$14^m.49$	$14^m.44$	35
	$14^m.0$								10
		$-0^m.05$	$-0^m.73$						38
$14^m.5$		$-0^m.04$	$-0^m.83$						38
	$13^m.02$	$-0^m.02$	$-0^m.72$						10
$16^m.3$									2
	$13^m.23$	$0^m.17$	$-0^m.72$						10
$13^m.4$									5
$13^m.6$									
$14^m.8$									
	$13^m.22$			$13^m.43$	$13^m.35$	$13^m.25$	$13^m.30$	$13^m.43$	35
	$14^m.80$	$-0^m.06$	$-0^m.96$						10
	$14^m.81$			$15^m.33$	$15^m.18$	$14^m.64$	$14^m.54$	$14^m.48$	35
$15^m.1$									2
$16^m.2$									2
	$15^m.25$			$15^m.77$	$15^m.60$	$15^m.11$	$15^m.15$	$15^m.65$	35
	$12^m.77$	$0^m.00$	$-0^m.78$						10
	$15^m.17$			$15^m.87$	$15^m.61$	$14^m.97$	$14^m.88$	$15^m.16$	35
	$14^m.41$	$0^m.26$	$-0^m.51$						10
	$14^m.68$	$0^m.16$	$-0^m.80$						44
	$14^m.0$								10
	$15^m.5$	$-0^m.045$	$-1^m.05$						44
	$15^m.67$			$15^m.60$	$15^m.63$	$15^m.65$	$15^m.77$	$16^m.12$	35
	$14^m.43$	$0^m.135$	$-0^m.77$						44
	$14^m.5$								10
$15^m.5$									47
	$14^m.96$	$+0^m.105$	$-0^m.815$						44
$16^m.6$	$16^m.9$	$(-0^m.34)$							5
	$16^m.90$			$17^m.44$	$17^m.17$	$16^m.83$	$16^m.78$	$16^m.85$	35

(continued)

No.	M_{pg}	V	$B-V$	$U-B$	I	R	G	B	U	References
	(10)	(11)	(12)	(13)	(14)	(15)	(16)	(17)	(18)	(19)
455	$16^{m}.0$									47
456		$16^{m}.21$	$0^{m}.31$	$-0^{m}.60$						10
		$16^{m}.15$			$16^{m}.40$	$16^{m}.31$	$16^{m}.15$	$16^{m}.24$	$16^{m}.55$	35
457		$14^{m}.35$	$0^{m}.20$	$-0^{m}.70$						10
		$14^{m}.33$			$14^{m}.70$	$14^{m}.56$	$14^{m}.26$	$14^{m}.32$	$14^{m}.60$	35
458		$15^{m}.54$			$15^{m}.18$	$15^{m}.25$	$15^{m}.78$	$16^{m}.05$	$16^{m}.79$	35
459		$15^{m}.65$	$1^{m}.93$							44
460		$16^{m}.50$	$0^{m}.42$	$-0^{m}.46$						3
		$16^{m}.47$			$16^{m}.32$	$16^{m}.32$	$16^{m}.57$	$16^{m}.73$	$17^{m}.17$	35
461	$15^{m}.5$									47
462		$17^{m}.2$	$0^{m}.1$	$-1^{m}.0$						10
463	$16^{m}.0$									47
464	$13^{m}.5$									47
465		$15^{m}.86$			$16^{m}.50$	$16^{m}.28$	$15^{m}.61$	$15^{m}.52$	$15^{m}.65$	35
466	$12^{m}.5$		$-0^{m}.32$	$-1^{m}.32$						38
467		$16^{m}.21$			$16^{m}.62$	$16^{m}.47$	$16^{m}.09$	$16^{m}.07$	$16^{m}.63$	35
468	$16^{m}.4$									2
469		$12^{m}.90$	$-0^{m}.28$	$-1^{m}.06$						10
470	$13^{m}.5$	$14^{m}.4$	$-1^{m}.$							5
	$14^{m}.3$									
471		$11^{m}.50$	$-0^{m}.30$	$-1^{m}.20$						10
472		$14^{m}.66$			$15^{m}.36$	$15^{m}.15$	$14^{m}.44$	$14^{m}.35$	$14^{m}.57$	35
473	$16^{m}.6$									2
		$17^{m}.02$			$16^{m}.65$	$16^{m}.75$	$17^{m}.21$	$17^{m}.53$	$18^{m}.07$	35
474	$14^{m}.5$	$15^{m}.11$	$0^{m}.09$	$-0^{m}.78$						5
		$15^{m}.09$	$0^{m}.12$							
		$15^{m}.09$			$15^{m}.41$	$15^{m}.30$	$15^{m}.02$	$15^{m}.06$	$15^{m}.07$	35
475		$13^{m}.10$	$0^{m}.20$	$-0^{m}.65$						10
		$12^{m}.98$			$13^{m}.46$	$13^{m}.29$	$12^{m}.86$	$12^{m}.85$	$13^{m}.36$	35
476		$13^{m}.82$	$0^{m}.03$	$-0^{m}.72$						10
477		$15^{m}.25$	$0^{m}.22$	$-0^{m}.58$						10
478										
479		$13^{m}.08$			$13^{m}.39$	$13^{m}.30$	$12^{m}.98$	$12^{m}.96$	$13^{m}.16$	
480		$12^{m}.90$	$0^{m}.15$	$-0^{m}.61$						10
		$12^{m}.89$			$13^{m}.42$	$13^{m}.23$	$12^{m}.77$	$12^{m}.76$	$13^{m}.20$	35
481		$14^{m}.54$			$15^{m}.37$	$15^{m}.04$	$14^{m}.28$	$14^{m}.05$	$13^{m}.78$	35
482		$14^{m}.54$			$15^{m}.37$	$15^{m}.04$	$14^{m}.28$	$14^{m}.05$	$13^{m}.78$	35
483		$16^{m}.09$			$16^{m}.47$	$16^{m}.35$	$15^{m}.98$	$16^{m}.00$	$16^{m}.52$	35
484	$15^{m}.7$									2
485		$14^{m}.6$								10
486		$14^{m}.7$								10
487		$13^{m}.05$	$0^{m}.07$	$-0^{m}.87$						10
488	$15^{m}.1$									5

White Dwarfs. Kinematics

No.	μ		θ	U, V, W (km s^{-1})	U', V', W' (km s^{-1})	V_r	V_r' (km s^{-1})	V_{tang} (km s^{-1})	π_{trig}	π_{ptm}	References
	(20)		(21)	(22)	(23)	(24)	(25)	(26)	(27)	(28)	(29)
1	0″.76		168°								5
2	0″.03		55°	34, −23, 18				45		0″.024	35
3			250°	−28, 32, −18		76, 82, 25					38
4	0″.55				−170, 4, −130	33, 79, −51					10
	−0″.37	−0″.41	220°		−116, 0, −75			138	0″.031	0″.019	35
								86			
5	0″.75										2
			210°	−28, 6, −33				44		0″.079	35
6	0″.48	−0″.22		93, −117, −51	42, −54, −24	11, 47, −88					10
									0″.0330 ± 0″.004		50
7	0″.04		77°	13, −5, 0				14		0″.013	35
	0″.036	0″.008		11, −4, 0		25, 62, −75					10
				16, 7, −14			69, 69, 19				9
8	0″.27		112°	51, −40, −30				72	0″.010	0″.018	35
								77?	0″.011		50
9	0″.11		103°								46
10	0″.29		135°	4, −14, −4				15		0″.091	35
11	+0″.44	0″.00		+102, −62, −7	+62, −37, −4	+21, +44, −87	108, +102, +57				10
	0″.38		204°	+114, −37, −56	+74, −12, −53			82	0″.026	0″.022	9
				−64, −38, −34					0″.028 ± 0″.005		50
12			130°	−11, −65, 13			−9, −6, −60				4, 47
13	0″.55		54°	106, −86, 89				163		0″.016	35
14	0″.32										5
15	0″.60		232°								5, 46
16	0″.20		231°	−38, −4, −7				39		0″.025	35
								39			22
	0″.25										5

(continued)

No.	μ		θ	U, V, W (km s^{-1})	U', V', W' (km s^{-1})	V_r	V_r' (km s^{-1})	V_{tang} (km s^{-1})	π_{trig}	π_{ptm}	References
	(20)		(21)	(22)	(23)	(24)	(25)	(26)	(27)	(28)	(29)
17	+1″260	−2″700			−6, −85, −67	+28, 46, −67			0″242		
									248		
									243		10
									223		
									233		
				3, −83, −70			54, 54, 3				9
	2″97		154°	−3, −60, −34				69	0″239	0″199	35
								56	0″23 ± 0″003		50
18	0″108		84°								
19											
20	0″56										2
21	0″00	−0″47	350°	−74, −113, −38	−29, −45, −16	+16, 22, −96					2
	0″47										46
22			80°								47
23	+0″27	+0″36		58, 16, 31	33, 9, 18	33, 42, −85	6				10
				61, 20, 22	36, 13, 9		61, 58, 10				9
	0″40		52°					27	0″045 ± 0″004	0″069	35
								38			
24											36
25	0″234		70°								44
26	0″182		77°								44
27	+0″15	0″00		48, −34, +4		54, 73, −43					10
			100°								47
28	+0″36	+0″27		94, −18, 63	58, −9, +31	50, 64, −58					10
			120°								47
29	0″00	−0″65		−42, −71, −77	−20, −34, −37	46, 51, −72					10
	0″66							50	0″060		5

(continued)

No.	μ		θ	U, V, W (km s^{-1})	U', V', W' (km s^{-1})	V_r	V'_r (km s^{-1})	V_{tang} (km s^{-1})	π_{trig}	π_{ptm}	References
	(20)		(21)	(22)	(23)	(24)	(25)	(26)	(27)	(28)	(29)
30	0″.51		160°	−13, −43, 1				45	0″.053		35, 5
31			115°				85, 87, 34				4, 46
32									0″.029 ± 0″.004		50
33			100°								46
									0″.017 ± 0″.003		50
34	+0″.43	0″.07		51, −55, 10	23, −25, 4	30, 12, −94					10
	0″.51		119°								46
35	0″.02		305°	−22, 10, 10				26		0″.027	35
36	0″.57	−0″.34		13, −30, −1	125, −288, −13	37, 20, −90			0″.061		10
	0″.66		122°	8, −20, −1				22	0″.061	0″.143	35
								51	0″.084 ± 0″.004		50
37	0″.20		134°								46
38	+0″.29	−0″.11			49, −64, −6	59, 52, −61					10
	0″.28							70	0″.018		5
39	−0″.25	−0″.12		−59, 22, −37	−29, 11, −19	59, 50, −63					10
40			292°			−2, 4, −1		5		(0″.087)	35
	0″.09										5
41	0″.02		100°	47, −38, 3				60	0″.014		35
42			5°	64, 66, 13							4, 47
43	+0″.04	−0″.09			−2, −31, −16	53, 34, −77					10
44	0″.36		95°	56, −55, 16				80		0″.014	35
45			300°				17, 10, −34				4, 47
46	+0″.421	−0″.121		40, −48, 6		67, 48, −57			0″.031 ± 0″.003		50, 10
				52, −39, −4			69, 66, 18				4
47	1″.15		116°	50, −64, −10			29, 29, −22	82	0″.064	0″.068	35
								85	0″.042?		

(continued)

No.	μ		θ	U, V, W (km s^{-1})	U', V', W' (km s^{-1})	V_r	V_r' (km s^{-1})	V_{tang} (km s^{-1})	π_{trig}	π_{ptm}	References
	(20)		(21)	(22)	(23)	(24)	(25)	(26)	(27)	(28)	(29)
									0″.059 ± 0″.005		50
48	0″.02		245°	−11, 7, −11				17		0″.041	35
49	0″.85	−0″.64			54, −75, −27	72, 62, −30					10
			127°					108	0″.037		35, 5
	1″.07				60, −84, −31			133	0″.049	0″.046	
									0″.052 ± 0″.004		50
50	0″.16	−0″.01		21, −20, 8		70, 71, −6					10
				−9, −51, −11			32, 36, 74				4
51	+0″.14	−0″.06			15, −24, +1	70, 41, −58					10
									0″.0216		50
52	0″.095	−0″.019		9, −13, 2		62, 19, −76					10
				−4, −17, 18			30, 23, −21				9
53											
54			70°	16, −14, 31	1, −7, 22		33, 21, −18				4, 47
55	0″.083	0″.010	83°	7, −5, +4		62, +16, −77					10
	0″.08									0″.042	35
									0″.011 ± 0″.004		50
56	0″.28		135°	28, −35, −18				48		0″.028	35
57	0″.36										2
58	0″.67										2
59	0″.67		98°								44
60	0″.54										2
									0″.037 ± 0″.003		50
61			115°								46
62			250°	−15, 8, −13				22		0″.017	35
63	0″.113		76°								44

(continued)

No.	μ		θ	U, V, W (km s⁻¹)	U′, V′, W′ (km s⁻¹)	V_r	V_r' (km s⁻¹)	V_{tang} (km s⁻¹)	π_{trig}	π_{ptm}	References
	(20)		(21)	(22)	(23)	(24)	(25)	(26)	(27)	(28)	(29)
64	0″08	−0″01		3, −5, +4		−20, −69, −69					10
	0″09		147°						0″070		48
65	0″012	−0″285			2, −18, −22	80, 18, −57					10
	0″31		180°								35
66			160°	81, −25, −10	−10, −13, −1		38, 33, −13				4, 47
67	0″88										2
68	0″43										2
			137°	75, −64, −20				101	0″026	0″020	35
								78			
69	0″71	0″36	63°	74, −29, 52		42, −38, −82					10
	0″83										46
70	0″36	0″31		54, −146, −7	25, −67, −3	88, 34, −33					10
				89, −133, −20	60, −54, −16		90, 83, 39				9
71			80°								46
72	0″47	−1″16			−4, −82, −32	87, 15, −47					10
									0″068		5
	1″25		162°	−8, −93, −43				102	0″072	0″050	35
								73			
73	0″173		33°								44
74	0″111		30°								44
75			120°								46
76	0″038	−0″052		6, −37, −7		90, 21, −38					10
	0″06		144	8, −51, −9				52	(0″008)	0″006	35
								40			
77	0″152	−0″008		41, −20, −2		84, 0, −53					10
				41, −20, −2			105, 92, (+69)				9
	0″16		94°	13, −24, 20				34	0″022	0″023	35

(continued)

No.	μ		θ	U, V, W (km s^{-1})	U', V', W' (km s^{-1})	V_r	V_r' (km s^{-1})	V_{tang} (km s^{-1})	π_{trig}	π_{ptm}	References
	(20)		(21)	(22)	(23)	(24)	(25)	(26)	(27)	(28)	(29)
								36			
78	0″.30	−1″.20			−9, −77, −34	90, 10, −42					10
	1″.24		166°						0″.015		35
79	0″.13	−0″.21			35, −17, −16	92, 18, −34					10
			149°		35, −17, −16		88, 78, (+52)	33	0″.036		9
	0″.25			3, −32, −8				34		0″.036	35
80	0″.116	−0″.023			40, −17, −2	90, +6, −42					10
	0″.12		101°	10, −25, 17				28	0″.023	0″.017	35
								32			
81	0″.070	−0″.076		35, −16, −15		91, 7, −41					10
	0″.12			35, −16, −15			80, 70, (+43)				9
			137°								46
82			170°								46
83	0″.051	−0″.086		32, −18, −20		89, −2, −46					10
				32, −18, −20			64, 51, (+27)				9
	0″.11		150°	2, −25, −4				25		0″.020	35
84	0″.090	−0″.030		38, −15, −7		91, 2, −42					10
85	2″.229	3″.420		−97, −12, −41							10
	4″.08		33°	56, 22, 60				85	0″.205	0″.223	35
								92	0″.207 ± 0″.002		50
86			150°								46
87	+0″.04	−0″.28		11, −65, −46	6, −35, −25	94, 28, −17					10
88	−0″.021	−0″.004		−5, 7, −10	−2, 3, −5	92, 1, −40					10
	0″.04		259°								46
89	0″.110		200°								44
90									0″.023 ± 0″.002		50
	0″.100	−0″.020		40, −15, −3		92, 1, −38					10

(continued)

No.	μ		θ	U, V, W (km s^{-1})	U', V', W' (km s^{-1})	V_r	V_r' (km s^{-1})	V_{tang} (km s^{-1})	π_{trig}	π_{ptm}	References
	(20)		(21)	(22)	(23)	(24)	(25)	(26)	(27)	(28)	(29)
	0″.11		92°								46
91	0″.26		210°								46
92									0″.023 ± 0″.0015		(50)
	0″.099	−0″.014		40, −14, −2		93, 1, −36					10
	0″.10		99°								46
93	1″.29	−1″.99			33, −50, −2	84, 52, 13					3
	2″.37							60	0″.179		5
	2″.37			33, −50, −10						0″.181	35
									0″.183 ± 0″.003		50
94	0″.102	−0″.034		40, −17, −3		94, 1, −34					10
	0″.11		108°	7, −29, 18				28	0″.023		35
95	0″.089	0″.001		40, −13, −3		92, −6, −39					10
				40, −13, −3			118, 104, (+79)				9
	0″.09		89°	7, −16, +19				26	0″.023	0″.018	35
								28			
96	0″.264	−0″.137			38, −16, 0	96, 13, −23					10
	0″.36							30	0″.062		5
97			165°	11, −39, −20				45		0″.016	35
98	0″.23	−1″.47		−150, −210, −70	−76, −106, −34	75, −35, −55					10
99	0″.10		111°	3, −20, 11				23	(0″.023)	0″.021	35
								28			
100	0″.085	0″.010		42, −13, 0		92, −10, −38					10
	0″.09		83°	7, −13, +20				25	0″.023 ± 0″.0015	0″.017	35
								28			
				42, −13, 0			105, 91, (+65)				9

(continued)

No.	μ	θ	U, V, W (km s^{-1})	U', V', W' (km s^{-1})	V_r	V_r' (km s^{-1})	V_{tang} (km s^{-1})	π_{trig}	π_{ptm}	References
	(20)	(21)	(22)	(23)	(24)	(25)	(26)	(27)	(28)	(29)
101	0″.054									2
			19, −52, +75				93		0″.026	35
										2
102	0″.36	127°								47
103		180°	67, −23, −37	60, −1, −18		109, 103, 58				4, 47
104	0″.30	227°	16, −2, −99				100		0″.014	35
105	0″.10	160°	5, −11, −4				13	0″.021	0″.036	35
							22			
			72, −50, −52			69, 67, (+34)				4
106		170°								47
107	0″.17 −0″.07			5, −36, 24	10, −73, 4			0″.0200 ± 0″.003		10, 50
108		285°	−1, 4, −4				211		0″.057	35
109	0″.44	141°	14, −209, 23				211	0″.003	0″.010	35
							0			36
110		120°	6, −21, 13				26		0″.027	35
										36
111	0″.27	21°	7, 8, 11				15		0″.089	35
							15			
112		160°	5, −35, −2			63, 49, 12				
113	−0″.03 +0″.29			−16, −45, −32	96, −26, 12					10
	0″.30						100	0″.014		5
114	0″.574 −2″.293			−42, −64, −17	84, −49, −25			0″.147		10
	2″.44	167°	−48, −73, −22				90	0″.155	0″.129	35
							75	0″.166		
115	1″.02	206°	−14, −20, −33				41	0″.122	0″.121	35
							40			

(continued)

No.	μ		θ	U, V, W (km s^{-1})	U', V', W' (km s^{-1})	V_r	V_r' (km s^{-1})	V_{tang} (km s^{-1})	π_{trig}	π_{ptm}	References
	(20)		(21)	(22)	(23)	(24)	(25)	(26)	(27)	(28)	(29)
								41			22
	1″07										5
116	0″26		77°	1, −4, 10				11		0″111	35
117	0″31										2
118	0″79		216°								46
119			215°	−1, −25, −45	−2, −10, −19		48, 37, −3				4, 47
120	−0″06	−0″40		−12, −51, −41		97, −23, 0					10
	0″36		189°								46
									0″028 ± 0″003		50
121	0″301		184°						0″040		
122	0″58										2
123	0″086		17°								44
124	0″050	−0″087			−8, −21, +1	93, −37, −1					10
125			170°								47
126	0″29										2
127			190°	37, −45, −31	24, −18, −11		67, 62, 16				4, 47
128						94, 15, 30					10
129	−0″502	−1″215		−14, 0, −8							10
									0″377 ± 0″003		50
130	−0″18	−0″93		13, −62, −40					0″062		10
									0″054		
									0″038		
									0″067		
	0″97		190°	17, −85, −53				102	0″058	0″047	35
								83	0″070 ± 0″003		50
131	0″49										2

(continued)

No.	μ		θ	U, V, W (km s^{-1})	U', V', W' (km s^{-1})	V_r	V_r' (km s^{-1})	V_{tang} (km s^{-1})	π_{trig}	π_{ptm}	References
	(20)		(21)	(22)	(23)	(24)	(25)	(26)	(27)	(28)	(29)
132			140°								47
133	0″.48		199°	38, −41, −30				63		0″.036	35
											36
134	0″.1										5
	0″.2		160°								47
135	0″.08	−0″.26		−65, −109, −23	−30, −51, −11	85, −52, +5					3, 10
	0″.27										5
136	0″.217		342°								44
137	−0″.29	−0″.31			17, −25, −50	94, 0, 34					10
138			220°								47
139											
140	0″.27		205°								38
	0″.2										5
141			120°	−30, −44, 9			25, 19, −26				4, 47
142	−0″.26	−1″.32			21, −56, −23	88, 16, 44					10
	1″.34		191°					65	0″.098		5
				18, −46, −19				53	0″.092	0″.120	35
								66	0″.089 ± 0″.003		50
143	0″.29		173°	12, −24, 1				27		0″.051	35
144			230°								47
145	0″.659		5°								44
146	−1″.030	0″.708		−5, −9, −18							10
147	1″.136	−0″.520			−32, −22, 28	59, −81, 4			0″.159		10
					−56, 11, 27		11, −7, −40		0″.102		9
	1″.25		115°	−33, −23, +28				49	0″.142	0″.121	35
								41	0″.148 ± 0″.007		50

(continued)

No.	μ		θ	U, V, W (km s^{-1})	U', V', W' (km s^{-1})	V_r	V_r' (km s^{-1})	V_{tang} (km s^{-1})	π_{trig}	π_{ptm}	References
	(20)		(21)	(22)	(23)	(24)	(25)	(26)	(27)	(28)	(29)
148	0″.064		337°								44
149			210°	13, −25, −21				35		0″.021	35
											36
150	0″.27		195°	14, −30, −11				35		0″.036	35
151	0″.055	−0″.018		−8, −3, 7		28, −95, −10					10
152	0″.49		185°	31, −68, −9				75		0″.031	35
153											5
	0″.51		194°	14, −76, −36				85	0″.029	0″.028	35
								83	0″.0310 ± 0″.004		50
	0″.49										2
154	1″.45	−1″.45			−184, 11, 62	−17, −93, −33					10
	2″.03		135°					60	0″.173 ± 30		5
155	−0″.13	−0″.20		−20, −21, −41	−10, −11, −20	59, −80, 12					10
				−17, −24, −41	−7, −14, −20		56, 38, 5				9
	0″.32		204°								46
156	0″.124		78°								44
157	0″.31		187°	9, −68, −21				72	0″.023	0″.021	35
								63			
158	0″.473		128°								44
159											
160			125°	−32, −41, 31				60		(0″.012)	35
161			210°	4, −10, −7				13		0″.024	35
162	−0″.20	−0″.32			28, −45, −48	84, −4, 55					10
					−25, −43, −82		−12, −14, −63		0″.0256 ± 0″.004		35, 50
163			220°	−6, −52, −59	−22, −24, −40		8, 5, −43				4, 47
164									0″.045 ± 0″.004		50

(continued)

No.	μ		θ	U, V, W (km s^{-1})	U', V', W' (km s^{-1})	V_r	V_r' (km s^{-1})	V_{tang} (km s^{-1})	π_{trig}	π_{ptm}	References
	(20)		(21)	(22)	(23)	(24)	(25)	(26)	(27)	(28)	(29)
165	−0″.034	−0″.011		36, −17, 0						10	
	0″.04		250°						0″.0062 ± 0″.		47
166	−0″.025	−0″.013			33, −19, 4						10
	0″.04		250°						0″.0062		47, 46, 50
167	−0″.036	−0″.014			37, −19, −2						10
	0″.04		250°						0″.0062		47, 46, 50
168	−1″.06	1″.31			77, 21, −16	26, −96, 10				0″.110	10
	1″.69				70, 47, −19		24, 8, −27			0″.105	9
			321°						0″.108 ± 0″.007		50
169	−0″.12	−0″.05		19, −8, −56		42, −87, 26					10
	0″.13		247°	14, −6, −40				43		0″.015	35
170			280°	13, 5, −16				21		0″.035	35
171	0″.278		267°								44
172			170°	−5, −85, −8			44, 40, −7				4
173					75, −102, −32	71, 32, 63					3
	0″.53							130	0″.019		5
			216°								46
174	−0″.40	−0″.00		94, −10, 0		74, −15, 64		70.5			3
	0″.34		271°	30, 3, −34				45	0″.047	0″.033	35
								35			
175			200°	31, −49, −15	21, −23, 0		69, 63, 18				4, 47
176	0″.60		210°	50, −53, −6				69		0″.039	35
177	1″.56		222°	47, −61, −31				83	0″.100	0″.088	35
								76	0″.099 ± 0″.003		50
								83			22
178	0″.03	−0″.28		−44, −58, −34		72, 4, 70					10
					−44, −58, −34		74, 72, (131)				9

(continued)

No.	μ	θ	U, V, W (km s^{-1})	U', V', W' (km s^{-1})	V_r	V_r' (km s^{-1})	V_{tang} (km s^{-1})	π_{trig}	π_{ptm}	References
	(20)	(21)	(22)	(23)	(24)	(25)	(26)	(27)	(28)	(29)
								0″.019		50
179	−0″.14 −0″.02			12, −2, −12	69, −11, 71					10
	0″.14	264°	24, −5, −25				36	0″.016	0″.019	35
							45			
180	0″.428	313°								44
181	0″.28	230°	20, −30, −27				45	0″.030	0″.029	35
							44			
182	0″.086	172°								44
183	−0″.34 0″.16			42, +1, −13	8, −98, 19					10
	0″.37		295°							5
184	0″.51									2
185	0″.00 0″.29		2, 36, −3	1, 21, −2	65, 4, 76		42			10
	0″.31	359°	2, 41, −5				54	0″.028	0″.035	35
186	0″.10 −0″.10			−27, −26, 17	63, −15, 76					10
187	0″.30									5
188	0″.180	138°								44
189	−0″.23 −0″.30		25, −78, −40	11, −36, −21	56, −28, 78					10
			66, −98, 17	52, −56, 36		124, 120, 76				4
								0″.042 ± 0″.004		50
190	0″.34									2
191	0″.072	344°								44
192	0″.34									5
193										
194	0″.16	139°	−21, −24, 25				40		0″.019	35
195	−0″.30 0″.20		122, 42, −36	42, 14, −12	40, −52, 75					3
196	0″.136	298°								44

(continued)

No.	μ		θ	U, V, W (km s⁻¹)	U′, V′, W′ (km s⁻¹)	V_r	V_r' (km s⁻¹)	V_{tang} (km s⁻¹)	π_{trig}	π_{ptm}	References
	(20)		(21)	(22)	(23)	(24)	(25)	(26)	(27)	(28)	(29)
197			290°								47
198	0″.38										2
			48°	−18, 21, 3				28		0″.065	35
199	0″.62										2
200	0″.1										5
	0″.2		215°								47
201	−0″.32	−0″.06		34, −16, −28		16, −76, 63					10
	0″.33		260°								47
202	0″.034	−0″.024		−3, −1, 16		52, 6, 85					10
					−21, −3, −13		17, 19, −34				9
203	0″.15	−0″.27		−120, −101, 1	−53, −43, 1	35, −41, 84					10
	0″.31		149°	−48, −38, 1				61	0″.024	0″.023	35
204	0″.278		267°								44
205			195°	−12, −39, −28				49		0″.015	35
206							41, 42, −10				9
207	0″.401		305°								44
208			270°	58, −18, 42			115, 115, 64				4
209	0″.36										2
210	−0″.80	0″.07		130, −34, −54	56, −15, −24	12, −69, 72					10
	0″.80		275°						0″.085 ± 0″.005		47, 50
211			217°	27, −90, −19				96	0″.024	0″.016	35, 50
	0″.33										5
212	−0″.19	−0″.17		26, −38, 2	14, −21, +1	49, 36, 80					10
	0″.27		223°	30, −36, 7	18, −18, 6		58, 64, 7				9, 47
									0″.024		50
213	−0″.43	0″.0		39, −16, −20		10, −65, 76					10

(continued)

No.	μ		θ	U, V, W ($km\,s^{-1}$)	U', V', W' ($km\,s^{-1}$)	V_r	V_r' ($km\,s^{-1}$)	V_{tang} ($km\,s^{-1}$)	π_{trig}	π_{ptm}	References
	(20)		(21)	(22)	(23)	(24)	(25)	(26)	(27)	(28)	(29)
	0".43		270°	40, −15, −18				46	0".046	0".046	35
214	0".108	−0".167		−74, −57, 15		34, −19, 92					10
	0".19		140°	−50, −30, 12				60		0".014	35
215	0".22		149°	−38, −39, 36				65		0".016	35
216	0".28										2
217	−0".50	0".25		193, 2, −17	103, 1, −9	8, −61, 79					10
	0".54		293°								5
218	−0".95	0".0			46, −19, −18	25, −25, 94			0".080		10
	1".07				49, −22, −37		62, 62, 11				9
			272°		56, −21, −21			63	0".071	0".075	35
								72	0".084		
219	0".119		111°								44
220			255°	28, −33, −4				44		0".016	35
	0".06										46
221	−0".078	−0".049		18, −26, −6		35, 3, 94					10
				38, −24, 50			110, 114, 60				9
			255°	19, −15, −7				25		0".013	35
222	0".163		275°								44
223	0".51										2
			261°								35
224	0".231		303°								44
225	−0".15	0".00		14, −16, −4	62, −28, −20	27, −9, 95					3
			260°	13, −9, −4				16		0".041	35
226	−0".054	0".044		12, 3, −1		15, −34, 93					10
227	2".66	−0".33			−55, 26, 8	−44, −90, −5				0".213	10
	2".68		97°						0".206 ± 0".8	0".189	44, 46

(continued)

No.	μ	θ	U, V, W (km s^{-1})	U', V', W' (km s^{-1})	V_r	V_r' (km s^{-1})	V_{tang} (km s^{-1})	π_{trig}	π_{ptm}	References
	(20)	(21)	(22)	(23)	(24)	(25)	(26)	(27)	(28)	(29)
228	0″.42									2
		201°	5, −38, 18				42		0″.045	35
229	−0″.11 −0″.26		−2, −44, −2		26, −5, +97					10, 3
								0″.033 ± 0″.004		50
			−7, −46, −21			58, 62, (78)				4
230	−0″.31 −0″.07			57, −37, −15	19, −14, 97					10
	0″.30	259°						0″.022 (50)		46
231		300°	55, 5, −18				58		0″.012	35
232	0″.049	277°								44
233	−0″.31 0″.02		90, −37, −19	41, −17, −9	11, −24, 96					10
	0″.37									5
234	0″.28	263°	66, −41, −7				78		0″.017	35
235	0″.63									2
236	−0″.096 0″.042	310°	31, −3, −5		17, 1, 98					10, 35
	0″.11									
			49, 10, −9				51		0″.010	
237	−0″.054 −0″.11			0, −5, −2	31, 31, 90					10
	0″.12	206°								47
238	0″.51	271°	85, −45, −14				97		0″.025	35
239	−0″.64 0″.28			113, −14, 4	−9, −42, 90					10
	0″.70	292°								46
240	−0″.16 −0″.10			23, −42, −3	15, 1, 99					10
	0″.26	234°								46, 47
241	0″.186	182°								44
242	−0″.124 0″.015		49, −22, −6	29, −13, −3	19, 18, 97					10
	0″.13	276°								46, 47

(continued)

No.	μ		θ	U, V, W (km s^{-1})	U', V', W' (km s^{-1})	V_r	V_r' (km s^{-1})	V_{tang} (km s^{-1})	π_{trig}	π_{ptm}	References
	(20)		(21)	(22)	(23)	(24)	(25)	(26)	(27)	(28)	(29)
243	0″.037	0″.028		−4, +11, −1	−8, 20, −1						10
	0″.04		53°	−3, 8, −1				1	(0″.023)	0″.021	35
244	0″.13	−0″.19		−45, −20, 17	−19, −8, 7	23, 27, 94					3
				−53, −30, −18	−27, −18, −28		14, 22, −37				4
			145°								47
245	0″.57		257°								44
246	−0″.38	0″.22		160, −9, 15		−10, −19, 98					10
				164, 0, −34			2, 6, −50				9
	0″.44		300°	95, −5, 9				96	(0″.023)	0″.021	35
								93			
247	0″.042		268°								44
248			320°								47
249	−0″.034	−0″.051		0, −57, 9		9, 16, 98					10
	0″.06										46
250			250°	19, −25, 4				32		0″.015	35
251	0″.126		262°								44
252	0″.00	−0″.20		−21, −31, −3		−7, −6, 99					3
				−22, −32, +7			67, 69, 10				4
			185°								47
253	−0″.29	0″.10		88, −24, 3	38, −10, 1	−2, +2, 99					10
	0″.28		305°								46
254	0″.187		100°								44
255	−0″.45	−0″.92			−5, −57, −21	−25, −32, 92					10
								60	0″.078 + 0″.011		5, 50
	1″.00								0″.082		
			208°		−4, −79, −28			84	0″.082	0″.057	35
256			30°								47

(continued)

No.	μ		θ	U, V, W (km s^{-1})	U', V', W' (km s^{-1})	V_r	V_r' (km s^{-1})	V_{tang} (km s^{-1})	π_{trig}	π_{ptm}	References
	(20)		(21)	(22)	(23)	(24)	(25)	(26)	(27)	(28)	(29)
257	0″.02	−0″.47		−236, −324, −44	−62, −79, 123	−30, −38, 87			0″.074		10
258	0″.005	−0″.008		−4, −2, 0		−4, 4, 99					10
259			320°								47
260	0″.35		280°	76, −35, 7				84		0″.020	35
261	0″.05										5
262	0″.32		137°								2, 47
263	−0″.149	−0″.088		10, −25, 2		−6, 8, 99		32			10
	0″.17			15, −28, 3				51	0″.016 ± 0″.003	0″.024	35
											50
264	−0″.50	0″.18		92, −29, 0	40, −13, 0	10, −75, 0					3
	0″.53		289°	85, −51, −67	33, −35, −67		−20, −9, −71				47
265	0″.499		268°								44
266	0″.50		297°	82, −21, −9				85		0″.028	35
267	−1″.077	−0″.462		56, −84, −9		−45, −39, 80			0″.048		10
									0″.069		
									0″.043		
	1″.30		249°	42, −59, −5				73	0″.067	0″.086	35
								92	0″.065 ± 0″.004		50
268	0″.58										2
			263°	62, −62, 14				89		0″.029	35
269			265°	11, −10, 3				15			35
270	0″.27										2
271	−3″.750	−1″.220			81, −114, 8	−38, −22, 90			0″.131		10
	3″.87		254°	81, −111, 8				139	0″.117	0″.131	35
								158	0″.135		
									0″.119		50

(continued)

No.	μ		θ	U, V, W (km s^{-1})	U', V', W' (km s^{-1})	V_r	V_r' (km s^{-1})	V_{tang} (km s^{-1})	π_{trig}	π_{ptm}	References
	(20)		(21)	(22)	(23)	(24)	(25)	(26)	(27)	(28)	(29)
272	−0″.086	−0″.026			13, −16, −1	−53, −47, 71					10
	0″.12		246°						0″.0186(50)		46, 47
273	0″.54		263°								44
274	−0″.40	−0″.02		35, −30, 10		32, 61, 72					10
	0″.39		266°	34, −30, 11				47	0″.027	0″.036	35
								67			
275	−0″.200	0″.220		162, 7, −37	68, 3, −16	17, 49, 86					10
	0″.30		317°	34, −1, −8				35		0″.037	35
276	0″.87							82	0″.048	0″.051	35
			265°	51, −56, 15				82	0″.048	0″.051	35
								87	0″.050 ± 0″.004		50
277											
278	0″.30										2
279	0″.06	−0″.22			0, −19, −19	−63, −55, 55					10
	0″.23		164°								46, 47
280	−0″.14	0″.00			33, −28, 11	−13, 24, 96			0″.0145		3
	0″.20										47, 50
281			250°	19, −36, 17				44		0″.025	35
			161°								47
282	0″.29		259°	53, −67, 16				87		0″.016	35
283	0″.107		268°								44
284	0″.00	−0″.25		−17, −30, −25	−12, −25, −26	−18, 25, 95	23, –, −28				3
				−17, −30, −24	−12, −25, −25		23, 36, −28				9
			175°								47
									0″.026 ± 0″.004		50
285	−0″.68	−0″.00			−87, −80, −38	−37, 4, 93					10
	0″.68		275°								47

(continued)

No.	μ		θ	U, V, W (km s^{-1})	U', V', W' (km s^{-1})	V_r	V_r' (km s^{-1})	V_{tang} (km s^{-1})	π_{trig}	π_{ptm}	References
	(20)		(21)	(22)	(23)	(24)	(25)	(26)	(27)	(28)	(29)
286	−0″.28	0″.00		92, −86, 42	51, −47, 23	−38, 4, 92					10
	0″.28		270°								46
287	0″.27		283°	133, −75, 54				162		0″.008	35
288							97, 108, 46				9
289			150°	−31, −9, −3				32		0″.011	35
290							51, 65, 0				9
291			230°								47
292	−0″.21	−0″.40		40, −15, −3		−57, −76, −33					10
	0″.45		208°								46
293	0″.32		269°	39, −40, 22				60		0″.024	35
	0″.30										5
294	0″.34										2
295			290°								47
296			295°	34, −14, 7				38		(0″.009)	35
297	−0″.84	−0″.41		96, −304, 80	52, −164, 43	−56, 4, 83					10
	0″.92		241°	62, −229, 52				243	0″.003	0″.018	35
								0			
298			260°	32, −50, 23				64		0″.011	35
299									0″.030 ± 0″.004		50
300	0″.33		297°	60, −38, 9				72		0″.021	35
301	−0″.15	0″.00		29, −34, 25	10, −12, 8	−73, −11, 68					38
			260°								47
302	0″.070										44
303			210°	−5, −7, 8						0″.035	35
304	−0″.13	0″.08		24, −7, 12	10, −3, 5	−32, 39, 86					3

(continued)

No.	μ		θ	U, V, W (km s⁻¹)	U′, V′, W′ (km s⁻¹)	V_r	V_r' (km s⁻¹)	V_{tang} (km s⁻¹)	π_{trig}	π_{ptm}	References
	(20)		(21)	(22)	(23)	(24)	(25)	(26)	(27)	(28)	(29)
				29, −14, −3	15, −10, −10		24, 39, −17				4
											38
305	0″.36		219°	−26, −61, 49				83	0″.012	0″.020	35
306	0″.319		217°								44
307	0″.214		244°								44
308			170°								47
309	−0″.050	−0″.126		−7, −25, −6	−4, −13, −3	−76, 3, 65					3
			200°	−7, −25, −7				27		0″.029	35
											38
310			150°	−26, −1, −3				26		0″.028	35
311			110°								47
312	0″.00	−0″.075		−16, −15, −5							3
				−23, −14, 4			63, 79, 12				4
			165°					23			35
											38
313							48, 64, (52)				9
									0″.010		50
314	−0″.432	−0″.220		16, −26, 5					0″.063		10
									0″.072 ± 0″.007		50
									0″.078		44
	0″.48		243°								46, 47
315	0″.1		350°					40			35
	0″.2										
316	−0″.04	−0″.11		−6, 27, −9		−85, +1, 53					3
									0″.019		50
317	0″.62			−103, −39, −17	−51, −19, −9	−34, 56, 75					10

(continued)

No.	μ		θ	U, V, W (km s^{-1})	U', V', W' (km s^{-1})	V_r	V_r' (km s^{-1})	V_{tang} (km s^{-1})	π_{trig}	π_{ptm}	References
	(20)		(21)	(22)	(23)	(24)	(25)	(26)	(27)	(28)	(29)
	0.″56		169°					74	0.″034		35
								76			
318							27, 45, −24				9
									0.″055 ± 0.″005		50
319	−0.″59	0.″0		58, −109, 115	29, −54, 57	−69, 34, 66					10
320											
321	0.″06	−0.″06		−17, −3, −32		−89, 3, 46					3
											38
322	−0.″09	−0.″024		−20, −123, −38	−90, −54, −17	−90, −54, −17					3
	0.″25		193°								46, 47
323	−0.″17	−0.″18		1, −29, 1	5, −117, 6	−91, −2, 41					10
	0.″25		223°								46, 47
324											
325	0.″078		96°								44
326	0.″30										2
327	0.″53		193°					73			35, 5
									0.″047		50
328	−0.″46	0.″74		131, 9, 63	56, 4, 27	−37, 62, 70					10
				141, −8, 43	66, −13, 7		24, 42, −28				9
	0.″86		322°					69	0.″059		35
								72	0.″063		
329	0.″066										10
330											
331	0.″42										2
	0.″52		137°								47
332	1.″61							125	0.″061		5

(continued)

No.	μ	θ	U, V, W (km s⁻¹)	U′, V′, W′ (km s⁻¹)	V_r	V_r' (km s⁻¹)	V_{tang} (km s⁻¹)	π_{trig}	π_{ptm}	References
	(20)	(21)	(22)	(23)	(24)	(25)	(26)	(27)	(28)	(29)
	1″.62	317°					115	0″.068		35, 50
								0″.74		
										5
333	0″.27									
334										
335	0″.59	215°					128			35
336	−0″.04 −0″.43		−79, −50, −5	−37, −24, −2	−43, 62, 66					10
	0″.62	188°						0″.034 ± 0″.004		47, 50
337	0″.00 −0″.72		−166, −157, −77	−74, −70, −34	−69, 44, 58					3
	0″.66	179°					92	0″.006		35
338		225°					24			35
										2
339	0″.32	340°					85			35
340		10°								47
341										
342		295°								47
343	0″.33	153°					20			35
								0″.0820 ± 0″.004		50
344	0″.00 −0″.61		−87, −55, −22	−41, −21, −6	−62, 55, 56					3
			−84, −58, −25	−38, −24, −9		65, −5				4
			−87, −55, −23	−41, −22, −7		61, 80, 10				9
	0″.56	176°					48	0″.042		35
							67			
345	0″.134	175°						0″.036		44
346	0″.43									2
								0″.057 ± 0″.005		50

(continued)

No.	μ		θ	U, V, W (km s⁻¹)	U′, V′, W′ (km s⁻¹)	V_r	V_r' (km s⁻¹)	V_{tang} (km s⁻¹)	π_{trig}	π_{ptm}	References
	(20)		(21)	(22)	(23)	(24)	(25)	(26)	(27)	(28)	(29)
347	0″.013	−0″.130			−32, −14, −9	−46, 68, 57					3
					−55, 20, 19		101, 121, 50				4
	0″.14		170°					44			35
									0″.0170		50
348	0″.27	−0″.24		−27, −9, −103	−13, −4, −46	−96, 13, 25					10
	0″.36										5
349			170°								47
350	0″.205		207°								44
351	0″.27		349°	67, −10, −1				68	0″.031	0″.019	35
								40			
352	0″.36		191°	40, −2, 14				42		0″.038	35
353											
354	0″.64		151°								44
355	−0″.42	−0″.31			−14, −56, 24	−85, 37, 37					10
	0″.50		232°						0″.0272 ± 0″.003		46, 47, 50
356	0″.110		9°								44
357	0″.27		335°								2, 47
358	0″.446		193°								44
359	0″.28										2
			187°	−27, −37, −16				48	0″.023	0″.028	35
								59			
360			345°				108, 128, 57				4, 47
361	0″.04	0″.08		4, 17, 1		−96, 23, 14			0″.025		50
											3
362	0″.112		148°								44
363	1″.67										2

(continued)

No.	μ	θ	U, V, W (km s^{-1})	U', V', W' (km s^{-1})	V_r	V_r' (km s^{-1})	V_{tang} (km s^{-1})	π_{trig}	π_{ptm}	References
	(20)	(21)	(22)	(23)	(24)	(25)	(26)	(27)	(28)	(29)
	1″.67	312°	47, −36, 44				74	0″.164	0″.111	35
							49			
							52			22
364	0″.00 −0″.10			−12, −12, −8	−80, 55, 30					4
	0″.10						20	0″.025		5
365	−1″.4 3″.5		(270), (−506), (32)	164, −303, −43	37, 79, 48					3
			286, −472, 53	180, −269, −22		−154, −145, −205				9
	3″.63	337°	191, −117, 47				229	0″.057	0″.073	35
	3″.59						302	0″.063		46
366		160°								47
367	0″.27	220°	−34, −27, +15				46		0″.026	35
							0			
368	0″.27	220°	−39, −26, 18				50		0″.024	35
369	0″.28	170°								2, 47
370	0″.28	21°						0″.025		2, 5, 47, 50
371		360°								47
372	−0″.41 −0″.46		−59, −181, 46	−26, −78, 20	−95, 31, 0					3
373	−0″.25 −0″.30		−33, −54, 14	−15, −24, 6	−82, 55, 13					4
			−47, −45, 16	−29, −15, 8		67, 84, 16				9
374	0″.00 −0″.29		−27, −52, −32	−13, −25, −15	−89, 44, 5					10
			19, −17, −34	33, −48, −17		0, 17, −51				9
	0″.27	191°						0″.035 ± 0″.004		46, 47, 50
375	0″.24 −0″.20		−4, −12, −58	−3, −9, −45	−93, 36, 0					3
376	0″.36	318°	20, −9, 25				33		0″.050	35
							33			22

(continued)

No.	μ		θ	U, V, W (km s^{-1})	U', V', W' (km s^{-1})	V_r	V_r' (km s^{-1})	V_{tang} (km s^{-1})	π_{trig}	π_{ptm}	References
	(20)		(21)	(22)	(23)	(24)	(25)	(26)	(27)	(28)	(29)
377	0.″46										2
			174°								47
378	0.″332		161°								44
379	−0.″28	−0.″29		−26, −62, 20	−14, −32, 10	−93, 37, −6					3
380			300°	15, 9, 34	9, 2, 14		46, 64, −5				4, 47
381	0.″04	0.″34		71, 28, 23	34, 14, 11	−42, 88, 23					10
	0.″34		7°								46
382							−12, 6, 63				9
	0.″27		42°	38, 40, −14				57		0.″022	35
383	0.″28										2
384	0.″101	0.″508			36, −7, 1	18, 89, 42			0.″47	0.″062	10
									0.″109		
	0.″54		110°	−2, 18, −37				41	0.″022	0.″063	35
								32	0.″081		
								41			22
385	0.″00	−0.″10		−11, −10, −8		−67, 74, 2					10
				−20, +1, −8			65, 83, 14		0.″0316		9, 50
386	0.″28										2
			1°					16	0.″084		5
			179°	−14, −3, −7				16	0.″084	0.″071	35
387	−0.″06	−0.″19			−4, −7, −1	−86, 48, −17			0.″118		10
	0.″20				−5, −8, 0			20			5
	0.″19		205°					9	0.″098	0.″100	35
								10	0.″102		
388											
389			90°	25, 22, −62				71		0.″010	35
390	0.″000	−0.″088		3, −18, −16	7, −15, −2	−65, 76, 0					3

(continued)

No.	μ		θ	U, V, W (km s^{-1})	U', V', W' (km s^{-1})	V_r	V_r' (km s^{-1})	V_{tang} (km s^{-1})	π_{trig}	π_{ptm}	References
	(20)		(21)	(22)	(23)	(24)	(25)	(26)	(27)	(28)	(29)
				3, −18, −5	7, −15, −2		35, 53, −16				9
392			200°								47
391	−0″.02	−0″.14			−22, −56, −17	−87, 42, −27					10
	0″.14			189°					0″.0107		46, 47, 50
393									0″.0560 + 0″.004		50
394	0″.02	−0″.20		−43, −10, −32		−31, 94, 12					10
				−40, −20, −34			40, 6, −11				9
	0″.20		174°	−34, −8, −25				43		0″.022	35
395	0″.00	−0″.26		−36, −29, −26	−22, −18, −16	−61, 79, −4					10
				−51, −9, −27	−37, 2, −17		77, 94, 26		0″.024		9, 50
396	0″.42		152°								44
397			360°	22, 12, 15				29		0″.025	35
398	0″.219		180°								44
399	−0″.46	−0″.70			−34, −40, 5	−74, 62, −26			0″.076		10
					−52, −24, 11		76, 91, 25				4
	0″.84		213°						0″.0830(50)		46, 47
400			45°	61, 13, −14						0″.011	35
401	0″.077	−0″.014		26, −23, −20		−53, 84, −14					10
			170°	−10, −8, −11				17		0″.024	35
402	0″.12	−0″.26			1, −18, −15	−84, 32, −45					10
	0″.29		155°								47
403	0″.27		204°	−137, −6, −19				138		0″.009	35
404			195°	−46, 8, −12				48		0″.015	35
	0″.13										46
405			20°	17, 7, 4				19		0″.019	35
								19			22

(continued)

No.	μ		θ	U, V, W (km s^{-1})	U', V', W' (km s^{-1})	V_r	V_r' (km s^{-1})	V_{tang} (km s^{-1})	π_{trig}	π_{ptm}	References
	(20)		(21)	(22)	(23)	(24)	(25)	(26)	(27)	(28)	(29)
406	−0″.31	−0″.64			−45, −41, −49	−64, 73, −26					10
	0″.71								0″.045		5
407	0″.511		144°								44
408			55°	32, 4, −11				34		0″.021	35
409			60°	29, 6, −13				32		0″.022	35
410	−0″.40	−0″.55		42, −18, 0		−38, 91, −16			0″.093		10
									0″.075		
									0″.052		
				47, −29, 2			38, 55, −13		0″.072		9
	0″.67		216°		−45, −19, 0			49, 50	0″.066		35
									0″.069		
									0″.071 ± 0″.002		50
411	0″.17	−0″.19		25, −22, −13		−68, −45, −58					10
412			175°	−38, −32, −31	−15, −16, −12		47, 63, −4				4
413	0″.32	−0″.09			8, −3, −13	−76, 36, −55					10
	0″.33		106°								46, 47
414	0″.247		138°								44
415	0″.044		131°								44
416	0″.63										2
			233°								47
417	0″.78	−0″.27		19, −9, 23		−62, 60, −50					3
	0″.82							60			5
			106°	32, −11, −53				63	0″.064	0″.064	35
								64	0″.068		
418			45°	36, 7, −3				37		0″.019	35
419											

(continued)

No.	μ		θ	U, V, W (km s⁻¹)	U′, V′, W′ (km s⁻¹)	V_r	V_r' (km s⁻¹)	V_{tang} (km s⁻¹)	π_{trig}	π_{ptm}	References
	(20)		(21)	(22)	(23)	(24)	(25)	(26)	(27)	(28)	(29)
420	0″.50							85	0″.028		5
			213°	−73, −19, −10				76	0″.028	0″.031	35
								83			
421	0″.21										2
			200°	−38, −22, −14				46	0″.025	0″.031	35
								59			
422	0″.39										2
			45°								47
423	0″.08	−0″.36		28, −34, 12	14, −16, 6	−56, −65, −52					10
	0″.37		167°								46, 47
424	0″.20		100°	31, −1, −48				57	0″.015	0″.017	35, 5
								64			
425			89°	18, 0, −18			94, 108, 43	26		0″.026	35
426	0″.49		159°						0″.0320(50)		2, 47
427	0″.450		114°								44
428			50°	66, −15, −8				68		0″.010	35
429	0″.27		333°	7, −2, 22				23		0″.055	35
430	0″.20	0″.22		72, −8, 4	35, −4, 2	11, 99, 6					10
	0″.34		56°	68, −7, −12				69	0″.029	0″.023	35, 50
								56	0″.058		
431	0″.045	−0″.310		−17, 14, −27	−9, 7, −14	32, 90, 28					10
				−30, −25, −39	−22, −32, −26		8, 20, −44				9
	0″.34		169°	−16, 14, −28				35	0″.045	0″.050	35
								33			
432	0″.420	0″.000			94, −55, −3	−48, 66, −57					10
					94, −55, −3		−81, −70, −3				9
											35

(continued)

No.	μ		θ	U, V, W (km s⁻¹)	U′, V′, W′ (km s⁻¹)	V_r	V_r' (km s⁻¹)	V_{tang} (km s⁻¹)	π_{trig}	π_{ptm}	References
	(20)		(21)	(22)	(23)	(24)	(25)	(26)	(27)	(28)	(29)
	0″.39		84°	58, 2, −47						0″.021	35
								75	0″.033		
433	0″.090	−0″.006		15, −2, −15		−51, 60, −61					10
				20, −8, −9			40, 50, −11				9
											38
	0″.20		80°								46, 47
434			285°								47
435	0″.31	0″.56		62, −40, 19	31, −20, 9	42, 82, 38					10
	0″.62		24°							0″.032	50
436	0″.27		67°								47
437	0″.35										2
438	−0″.24	−0″.66		−77, −38, −36	−44, −21, −20	−24, 88, −40					10
	0″.73							50	0″.074		5
	0″.71		200°		−32, −15, −15			38	0″.074	0″.090	35
								46			
439											
440	0″.31	−0″.16		33, −32, −50		−49, 55, −67					10
				48, −49, −29			20, 28, −31				9
	0″.30		121°	32, −38, −56				75	0″.011	0″.019	35
								72	0″.052		
441	0″.28		102°								2, 47
442	0″.51										2
443			35°	35, 3, 9				36		0″.019	35
444	0″.00	−0″.33		−18, −25, −17	−11, −16, −11	−39, 68, −61					10
									0″.042 ± 0″.003		50
445			85°	23, −3, −16				28		0″.013	35
446	0″.000	−0″.275		−25, −40, −23	−9, −15, −9	−42, 64, −65					10

(continued)

No.	μ		θ	U, V, W (km s^{-1})	U', V', W' (km s^{-1})	V_r	V_r' (km s^{-1})	V_{tang} (km s^{-1})	π_{trig}	π_{ptm}	References
	(20)		(21)	(22)	(23)	(24)	(25)	(26)	(27)	(28)	(29)
				−51, 0, −63	−35, +25, −49		113, 122, 62				9
	0″.26		181°								46, 47
447	0″.41		192°						0″.069 ± 12		44
448	0″.13	−0″.18		21, −46, −14		−62, −4, −78					10
									0″.064 ± 0″.012		50
	0″.22		144°								46, 47
449	0″.117		265°								44
450	0″.43		44°	57, 17, 2				60	0″.040	0″.034	35
451	0″.651		161°								44
452	0″.21	−0″.02		51, −17, −30		−51, 10, −85					10
	0″.21		94°	54, −17, −24			44, 46, −7				9, 46, 47
453	0″.20		96°								44
454			240°				29, 37, −22				4, 47
455	0″.29		266°	−128, 28, −34				135		0″.010	35
456			80°								47
457	0″.16	−0″.30		−119, −80, −23	−43, −16, −9	−25, 59, −77					10
				−128, −60, −50	−52, 4, −36		85, 91, 34				9
	0″.34		209°	−61, −39, −12				73		0″.022	35
458	0″.51	0″.04		65, −17, −29	46, −12, −21	−1, 85, −53					10
			83°	55, −13, −22				61	0″.050	0″.041	35
	0″.52							50			
459											
									0″.048 ± 0″.005		50
	1″.26		82°	92, −23, −33				100	0″.049	0″.061	35
	1″.36							123	0″.043		47
460											
461	0″.520	−0″.045			98, −48, −40	−32, 51, −80			0″.061		3

(continued)

No.	μ		θ	U, V, W (km s^{-1})	U', V', W' (km s^{-1})	V_r	V_r' (km s^{-1})	V_{tang} (km s^{-1})	π_{trig}	π_{ptm}	References
	(20)		(21)	(22)	(23)	(24)	(25)	(26)	(27)	(28)	(29)
									0″048		
									0″043		
	0″59		91°		79, −28, −41			93	0″022(50)	0″030	35
462			90°				65, 69, 14				4, 47
463	−0″20	−0″24				−11, 69, −71					10
464			80°				60, 67, 9				4, 47
465			115°								47
466			75°	76, −22, −12				80		0″009	35
467			90°								47
468			200°	−34, 8, −39				52		0″014	35
469	0″61										2
									0″038		50
470	−0″054	−0″011		−8, 1, 2		−13, 55, −83					10
				−6, −5, +12			40, 44, −11				9
471	0″26										
			80°								5, 47
472	0″003	−0″003		1, −2, −1		−14, 49, −86					10
473			260°	−22, 10, 4				24		0″015	35
474	0″29			−2, −24, −37							2
			154°					44		0″030	35
475	0″07		210°	−11, −4, −5				13		0″029	35, 5
476	−0″47	−0″30		−63, −4, −1	−35, −2, −1	−2, 62, −79				0″054	10
	0″56		237°	−49, −3, −1				39			
477							88, 96, 37				9
	0″30		116°								47
478	0″47	0″02		127, −51, −37	62, −25, −18	17, 82, −54					10

(continued)

No.	μ		θ	U, V, W ($\mathrm{km\,s^{-1}}$)	U', V', W' ($\mathrm{km\,s^{-1}}$)	V_r	V_r' ($\mathrm{km\,s^{-1}}$)	V_{tang} ($\mathrm{km\,s^{-1}}$)	π_{trig}	π_{ptm}	References
	(20)		(21)	(22)	(23)	(24)	(25)	(26)	(27)	(28)	(29)
	0″.42		87°								46
479											
480										(0″.065)	35
481	−0″.23	−0″.06			−13, 4, 1	25, 84, −47					10
					−21, −24, 17		17, 23, −34				9
	0″.27		254°		−22, 7, 0			23	0″.056	0″.058	35
									0″.051 ± 0″.002		50
482										0″.011	35
483	0″.36		78°		106, −32, −11			111		0″.016	35
484	0″.57										2
485	−0″.30	−0″.40			−54, −29, 13	−24, 1, −97					10
	0″.50		217°								46
486	0″.02	−0″.60			−33, −81, 13	−26, −4, −96					10
	0″.68		178°								46
487	0″.60	−0″.67	138°		10, −35, 3	−28, −15, −95					10, 47
	0″.90								0″.120 ± 0″.007		46, 50
488	0″.60		157°								5, 47

References

[1] Hintzen, P., Strittmatter, P. A., Burbidge, E. M. and Wickramasinghe, D.: 1975, *Astrophys. J.* **202,** 191.
[2] Hintzen, P. and Strittmatter, P. A.: 1974, *Astrophys. J.* **193,** L111.
[3] Eggen, O. J. and Greenstein, J. L.: 1967, *Astrophys. J.* **150,** 927.
[4] Trimble, V. L. and Greenstein, J. L.: 1972, *Astrophys. J.* **177,** 441.
[5] Nesterenko, N. M. and Chuprina, V. P.: 1976, *Soobscheniya SAO* **17.**
[6] Eggen, O. J.: 1968, *Astrophys. J. Suppl.* **143,** 97.
[7] Weidemann, V.: 1966, *J. Quant. Spectroscop. and Rad. Transf.* **6,** 691.
[8] Shipman, H. L.: 1971, *Astrophys. J.* **166,** 587.
[9] Greenstein, J. L. and Trimble, V. L.: 1967, *Astrophys. J.* **149,** 283.
[10] Eggen, O. J. and Greenstein, J. L.: 1965, *Astrophys. J.* **141,** 83.
[11] Giclas, H. L., Burnham, R. and Thomas, N. G.: 1962, *Lowell Obs. Bull.* No. **118.**
[12] Luyten, W. J.: 1962, *A Search for Faint Blue Stars* No. **31.**
[13] Feige, J.: 1959, *Astrophys. J.* **129,** 600.
[14] Çhawra, E.: 1959, *Bull. Tonantzintla* No. **18.**
[15] Weidemann, V.: 1960, *Astrophys. J.* **131,** 638.
[16] Thackeray, A. D.: 1961, *Monthly Notices Roy. Astron. Soc. S* South Africa **20,** 40.
[17] Luyten, W. J. and Herbig, G. H.: 1960, *Harvard Announcement Card.* **1474.**
[18] Hardie, R. H. and Heiser, A. M.: 1966, *Publ. Astron. Soc. Pacific* **78,** 171.
[19] Kamp, P. van de, Iang, C. I. and Worth, M. D.: 1966, *Publ. Astron. Soc. Pacific* **78,** 167.
[20] Eggen, O. J. and Stoy, R. H.: 1961, *Roy. Obs. Bull.* No. **24.**
[21] Bell, R. A.: 1962, *Observatory* **82,** 68.
[22] Angel, J. R. P.: 1977, *Astrophys. J.* **216,**
[23] Gurzadyan, G. A.: 1972, *Astrophysica* **8,** 479.
[24] Stephenson, C. B.: 1960, *Publ. Astron. Soc. Pacific* **72,** 387.
[25] Burbidge, E. M. and Strittmatter, P. A.: 1971, *Astrophys. J.* **167,**
[26] Greenstein, J. L.: 1974, *Astrophys. J.* **189,** L131.
[27] Greenstein, J. L.: 1974, private communication.
[28] Routly, P. M.: 1972, *Publ. U.S. Naval Obs.* **20,** 6.
[29] Angel, J. R. P., Hintzen, P., Strittmatter, P. A. and Martin, P. G.: 1974, *Astrophys. J.* **190,** L71.
[30] Greenstein, J. L.: 1957, *Astrophys. J.* **126,** 23.
[31] Kraft, R.: 1964, *Astrophys. J.* **139,** 457.
[32] Humason, M. L.: 1938, *Astrophys. J.* **88,** 228.
[33] Luyten, W. J.: 1952, *Astrophys. J.* **116,** 283.
[34] Greenstein, J. L.: 1976, *Astron. J.* **81,** No. 5.
[35] Greenstein, J. L.: 1975, *Astrophys. J.* **196,** L117.
[36] Eggen, O. J. and Greenstein, J. L.: 1965, *Astrophys. J.* **142,** 925.
[37] Greenstein, J. L.: 1969, *Astrophys. J.* **158,** 281.
[38] Weidemann, V.: 1977, *Astron. Astrophys.* **59,** 411.
[39] Fusi-Pecci, F. and Renzini, A.: 1976, *Astron. Astrophys.* **46,** 447.
[40] Mengel, J. G.: 1976, *Astron. Astrophys.* **48,** 83.
[41] Sweeney, M. A.: 1976, *Astron. Astrophys.* **49,** 375.
[42] Schulz, H.: 1976, Paper presented at Second European Workshop on White Dwarfs, Rome.
[43] Wickramasinghe, D. T. and Bessell, M. S.: 1977, *Monthly Notices Roy. Astron. Soc.* **181,** 713.
[44] Greenstein, J. L.: 1974, *Astron. J.* **79,** 964.
[45] Luyten, W. J.: 1964, Publ. of the Astron. Obs. Univ. of Minnesota, 3, No. 14.
[46] Giclas, H. L., Burnham, R. and Thomas, N. G.: 1962: *Lowell Obs. Bull.* No. 118.
[47] Oke, J. B. and Shipman, H.: 1970, *IAU Symp.* **42,** held at St. Andrews, Fife, Scotland.
[48] Gershbergh, P. E.: 1978, *Vspykhivauushiye Zvezdy Malykh Mass*, Nauka, Moscow.
[49] Harry, L. and Shipman, H.: 1979, *Astrophys. J.* **228,** 240.
[50] Eggen, O. J.: 1979, *Astrophys. J. Suppl.* **39,** No. 1.
[51] Lacy, H. C.: 1977, *Astrophys. J.* **218,** 444.
[52] Green, R. F., Richstone, D. O. and Schmidt, A.: 1978, *Astrophys. J.* **224,** 892.
[53] Cowley, A. P., Crampton, D. and Hesser, J. E.: 1977, *Publ. Astron. Soc. Pacific* **89,** 531.
[54] *Obschiy Catalog Peremennykh Zvezd*, III, 1971.
[55] Svechnikov, M. A.: 1969, *Catalog orbit. elementov, mass, svetimostey.*
[56] Wegner, G.: 1973, *Monthly Notices Roy. Astron. Soc.* **163,** 381.
[57] Liebert, J., Conrad, C., Dahu, Gresham, M., Herge, E. U., Moore, R. L., Romanishin, W. and Strittmatter, P. A.: 1979: *Astrophys. J.* **229,** 196.

GPSR Compliance

The European Union's (EU) General Product Safety Regulation (GPSR) is a set of rules that requires consumer products to be safe and our obligations to ensure this.

If you have any concerns about our products, you can contact us on ProductSafety@springernature.com

In case Publisher is established outside the EU, the EU authorized representative is:

Springer Nature Customer Service Center GmbH
Europaplatz 3
69115 Heidelberg, Germany

Batch number: 10620141

Printed by Printforce, the Netherlands